Patrick Moore's Practical Astronomy Series

For other titles published in this series, go to
www.springer.com/series/3192

So You Want a Meade LX Telescope!

How to Select and Use the LX200 and Other High-End Models

Lawrence Harris

Lawrence Harris
Stowupland
UK
Lawrence@astronomer.plus.com

ISSN 1431-9756
ISBN 978-1-4419-1774-4 e-ISBN 978-1-4419-1775-1
DOI 10.1007/978-1-4419-1775-1
Springer New York Dordrecht Heidelberg London

Library of Congress Control Number: 2010925109

Printed on acid-free paper

Springer is part of Springer Science+Business Media (www.springer.com)

Acknowledgments

Books take time to be written. During that time, various people are called upon to help resolve the puzzles and queries that arise. Many people helped me arrive at the end of this project, so I want to acknowledge that help here.

I am indebted to Richard (Dick) Seymour, Andrew Johansen, Mike Weasner, and Jason Ware for their sterling work, reading drafts of the components of Chap. 2 and offering several useful suggestions based on their own expertise of the LX scope series. Their help during the writing of the PEC section was unsurpassed.

Rather than merely using my own images I posted a request for contributions and was delighted with the quality of images supplied to me. I regret that only a few could be selected because of space limitations. My thanks to all who responded.

My grateful thanks to Ray Gralak, the author of *PEMPro* who kindly agreed to check the draft of the PEC software section for errors. Terry Platt helpfully checked Chap. 9 and provided some interesting pictures.

Dr. R.A. Greiner and Dr. Clay are in a class of their own when it comes to expertise of the mechanics and operations of the LX200/LX400 series of telescopes. Both are credited elsewhere, but must be thanked here for their valuable contributions.

John Watson and Maury Solomon of Springer have been instrumental in helping me to get this book suitably prepared for publication. Their patience with my endless queries has lasted a long time and I am very grateful.

My better half, Marion has read this book as a nonastronomer to check that I have not been throwing astronomical acronyms around too excessively. Apart from that, never have we drunk so much tea and coffee each day as during the final run toward chapter submission. The book would not have happened had she not been more than patient with me.

Any errors that get through, however, are mine.

Contents

About the Author

Lawrence Harris retired from his job as a British government space scientist following nearly two decades at the leading edge of space research. For almost 20 years, Harris has been a columnist writing about weather satellites and space research in some of the top astronomical publications. He has also contributed extensively to astronomical society magazines. Harris is the proud owner of several Meade LX telescopes, including the top end LX400 telescope.

Chapter 1

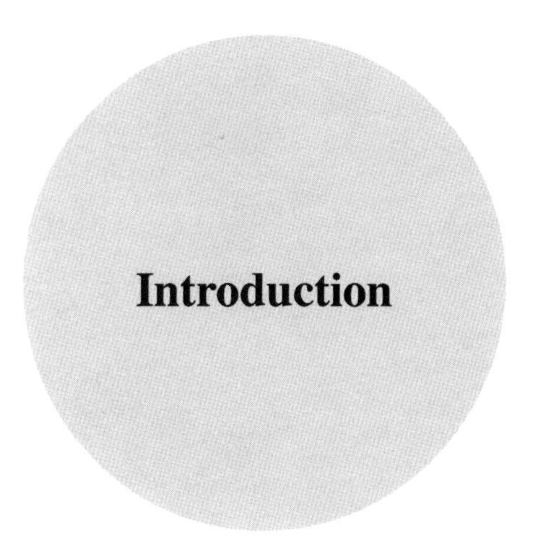

Introduction

Computers and Astronomy

Perhaps every generation of astronomers believes that their telescopes are the best that have ever been. They are surely all correct! The great leap of our time is that computer-designed and machined parts have led to more accurately made components that give the astronomer ever better views. The manual skills of the craftsman mirror grinder have been transformed into the new-age skills of the programmer and the machine maker. (The new products did not end the work of craftsman telescope makers, though. Many highly skilled amateur/professional opticians continued to produce good-quality mirrors that are still seen today.) Amateur-priced telescopes are now capable of highly accurate tracking and computer control that were once only the province of professionals. This has greatly increased the possibilities of serious astronomy projects for which tailor-made software has been developed. Add a CCD camera to these improved telescopes (see Chap. 3), and you bring a whole new dimension to your astronomy (see Fig. 1.1).

Look Before You Leap!

But first, a word of caution. Unless you are already familiar with astronomy and basic telescopes, it is not wise to start spending large amounts of money on a well-featured telescope. Such an instrument might otherwise be subsequently abandoned due to a perceived overcomplexity coupled with a waning interest. For absolute beginners, many seasoned amateurs would recommend that the first purchase of an instrument

L. Harris, *So You Want a Meade LX Telescope!*, Patrick Moore's Practical Astronomy Series,
DOI 10.1007/978-1-4419-1775-1_1, © Springer Science+Business Media, LLC 2010

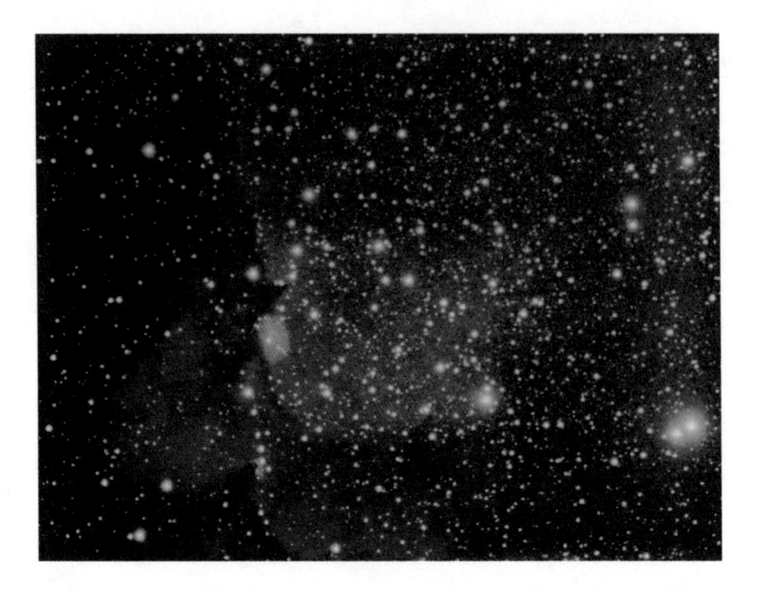

Fig. 1.1 NGC7380 100 min total exposure using a 12-in. (30-cm) Meade LX400 telescope, Starlight Xpress SXV-H9C one-shot color camera, and Starlight Xpress Adaptive Optics unit (image credit: Lawrence Harris)

for astronomy should be a pair of binoculars. A pair of 10×50 binoculars (*10* refers to the magnification, and *50* refers to the aperture diameter in millimeters) provides an excellent instrument for an introduction to the hobby. Binoculars are totally portable, so they can be taken into your back garden or to darker skies out of town. Here, they will give outstanding views, showing significantly more than the naked eye, and offer the best visual introduction to the night skies. They should reveal much fainter and often more colorful stars as well as the steady shining light of planets, star clusters, and even an asteroid or two. A vague patch of light resolves, becoming the Andromeda Galaxy; the Pleiades or other well-known star clusters all take on a whole new appearance. Your familiarity with the sky is likely to improve to the point where you can identify the brighter stars and several planets. Binoculars offer the chance to test your interest in the night sky without spending anything more than about $80. A good star atlas is also helpful!

Advancing from Binoculars

Having had your appetite whetted by at least a trial period with a pair of binoculars, you may well decide to move on and buy an actual telescope. Again, a word of caution here! A decent small telescope – see Fig. 1.2 – that can enable you to see fainter deep space objects and provides a manually controllable mount may cost about $250. This is surely preferable to the possibility of spending over $4,000 on

Fig. 1.2 A small refracting telescope on a low-cost mount. This telescope includes a small finder telescope, a right-angled prism, two low-cost eyepieces, a weight balance arm, and manual drives

an instrument that could fail to maintain your interest. An excellent way forward is to chat about your ideas with your local astronomical society. There will always be members that have significant experience using medium-sized telescopes, if not larger ones, and such societies may even be able to offer suitable telescopes on temporary loan. At the very least, you should be able to join others using a well set up scope and to experience its capabilities.

Telescope Types

No matter what price range you look at, any small telescope will be one of only a few types: the refractor, the reflector, and various forms of the Schmidt-Cassegrain telescope (usually referred to as an SCT) that uses an optical configuration that is a combination of a main mirror, secondary mirror, and correcting plate. Another advanced form, the Ritchey–Chrétien design, combines mirrors and lenses to eliminate many of the major optical imperfections often experienced with lesser optics.

Small Refractors

The refractor essentially uses two lenses that refract (bend) incoming light rays, bringing them to a real focus near the eyepiece. Refractors often have long tubes due to the focal length of the main objective lens and therefore can give the false impression of being powerful. Focal length (and consequent magnification) is often associated with power, although this is quite misleading. The most important parameter of a telescope is its aperture – the diameter of its objective (in the case of a refractor) or the diameter of the mirror (in the case of a reflector). It is this dimension that defines the instrument's light grasp. The larger the aperture, the greater the light grasp. Because the light-gathering power is related to the actual area that collects the incoming light, the telescope power changes as the square of the diameter, so a 30-cm-diameter telescope has four times the light grasp of a 15-cm telescope. This desire to get ever larger telescopes is sometimes referred to as aperture fever!

Refractors can become very costly as the aperture increases, mainly due to the increasing cost of manufacturing high-quality optical components for the main lens. Keen amateur astronomers often use refractors for planetary imaging, although a good-sized reflector can provide an image of comparable quality. Viewing through a high-quality refractor on a good mount can produce an excellent image. A well-aligned, sturdy, good-quality mount is an important factor in obtaining the best viewing conditions with any telescope. Spindly, wobbly mounts are liable to be buffeted in the slightest breeze, causing multiple vibrations to any image and really spoiling the viewing experience. The main limitation in image quality with some refractors is the presence of chromatic (color) fringes or other aberrations. The components of the main lens have to be extremely well matched to fully eliminate, or at least minimize, such aberrations.

Buying a refractor with a basic mount in the $200+ price range should get you a modest telescope of about 5 or 6 cm diameter that should be a significant improvement to your observational experience. Setting up this type of mount is usually accomplished within about 10 min, and a small collection of useful accessories of nominal quality is invariably provided.

Refractor Accessories

The accessories box usually includes a right-angled prism (this acts as a diagonal mirror) – see Fig. 1.3 – that slots into the near end of the telescope where the eyepiece would normally go. The observer can then view in a more comfortable position rather than contorting to view through the eyepiece at the rear end of the telescope – especially when viewing something high in the sky. This particular box includes a small finder and three eyepieces. Such a finder, designed to fit on the tube, usually comes with even the smallest scopes and should be adjustable for alignment with the main tube. A Barlow lens (used to increase the magnification of an eyepiece) may also be included. It should be understood that at this level, the optics of each component are probably adequate but unlikely to be sparkling. It is likely to be an introductory scope with some basic facilities to make using it fairly easy.

Fig. 1.3 The accessories that come with a small telescope usually include a right-angled prism. This enables the astronomer to view the heavens in a comfortable position rather than getting a cricked neck

The telescope manual should describe basic astronomical principles and how they apply to the setting up and alignment of your telescope; it might even include simple polar alignment. By alignment, we are referring to the need to set the telescope's main rotation axis parallel to Earth's own polar axis so that the telescope can be driven by motor to keep pace with Earth's rotation. The main operational difficulty likely to be found with this type of low-cost scope is the fairly small field of view. Although giving good views of the Moon, any attempt at finding galaxies or other small targets can be daunting; however, it does enable you to discover whether you want to move on with your interest in practical astronomy or settle down to a life in the armchair with a book instead. Do seek advice and suggestions from your local astronomical community.

If your interest in active astronomy develops, a basic telescope can quickly lose its appeal, and you may soon decide to upgrade. You can probably sell such a scope via various specialist Web forums or via your local society. By the time you have developed a serious interest in observing, you can decide on the option of simply spending more cash on a much better scope or trying out better scopes via your local astronomical society with a view to possible later purchase. Inevitably you are likely to appreciate the potential of a better scope, and aperture fever can set in! Discuss your thoughts and ideas with fellow astronomers. This might be at the local astronomical society meeting or better still at a star party. If your luck is really in you might get a chance to see and try a good amateur telescope owned by an enthusiast. This person will be able to tell you far more than a chat with a dealer might reveal. Many specialist dealers are also keen astronomers, and you might meet up with them at a star party.

Reflectors

Aperture for aperture, reflectors are usually cheaper than refractors. The most common design is the *Newtonian* reflector; this incorporates a small secondary plane mirror that reflects light from the primary mirror, off-axis into an eyepiece. For apertures of above 10 cm, the cost of producing a quality parabolic mirror (the shape of a properly designed telescope mirror) is far less than that of producing a quality combination achromatic lens. The cost of an accurately ground 25-cm mirror is but a fraction of the cost of an equal-sized good-quality objective. Because of their relative cheapness, amateurs tend to move to reflectors when they decide to buy a better and larger telescope. As with many experiences in astronomy, the better the mount, the better the resultant image is likely to be.

You may decide to advance directly from binoculars to a medium-sized scope, in which case you are likely to be told by other amateurs to consider buying a 15-cm (6-in.) reflector – see Fig. 1.4. This is often considered to be an ideal size for a first, more capable, telescope: the aperture is sufficient to see quite deeply into space, yet the scope is light enough not to require heavy, expensive mounts. The term *deep space* is used to refer to objects at increasingly large distances from our own planet, often extragalactic objects that are exceedingly faint. With such a telescope the heavens open up to you. The planets lose their mysteriously taunting tiny spherical appearance, as revealed in lesser instruments, and show detail – probably more than you may realize at first glance. Hundreds of asteroids become identifiable; lunar observations take on a far more serious meaning, and far fainter galaxies come within your grasp.

Also coming into the range of medium-sized scopes is the 8-in. (20-cm) reflector, especially in the form of the SCT (mentioned earlier). Because of the considerably greater facilities offered by these scopes, they are priced toward the higher end of the price range of medium-sized scopes.

Moving On Up

So far we have introduced a possible sequence of events; it starts with getting your first pair of binoculars, on to a basic telescope, and to realizing that you want to move on to an even better observing experience. A glance at the brochures and Web sites (see below) of various telescope suppliers reveals useful information. First, there are many individual telescopes and mounts in each price range, all catering to the beginner, the enthusiast, and the advanced amateur. Second, you may find that a specific telescope and mount combination is often available at marginally different prices from different vendors. They may charge according to the after-sales service that they offer; consequently, a general store price may be cheaper than a specialist outlet, but you may wish to inquire about possible technical support after purchase!

Fig. 1.4 Bresser Messier N160 6-in. (15-cm) reflector. This is a very popular size, combining portability with observing capability (image courtesy: Telescope House, UK)

Telescope Suppliers' Web Sites

Some United Kingdom Suppliers

http://www.iankingimaging.com/
http://www.scsastro.co.uk/
http://www.sherwoods-photo.com/
http://www.telescopehouse.com/
http://www.telescopesales.co.uk/

Some US Telescope Suppliers

http://www.celestron.com
http://www.meade.com/dealerlocator/index.html
http://www.skywatcherusa.com/

Enter the Goto Handbox

Many telescopes costing more than about $900 now include a *goto* facility. This involves a small computer often located in a handbox, or a keypad, plugged into the telescope's mount – usually via a standard connector. The manual tells you how to power up the mount, and then the handbox asks for the scope to be aligned, as a basic requirement. Depending on the capabilities (and therefore price) of the system, you have to enter some basic information that might include time (unless there is a built-in clock to remember it), whether you are using UTC or local time, and your longitude and latitude. Once these parameters have been entered, the next stages for the user will involve a polar alignment followed by a star alignment using one, two, or three stars. The experienced amateur soon masters these tasks, and for advanced larger telescope users, there is software available to aid both star alignment and polar alignment. More advanced telescopes often require less start-up intervention, as we shall see. The LX400 is first powered up, then its self-identification GPS location is measured, and it then stores the position where it was left pointing before power-off. Consequently, it is virtually ready to go within a minute or two of power on.

Star and Polar Alignment

At the beginning of a session when you start aligning this type of telescope mount, you have to accurately point the telescope at one or more known (and easily recognizable) stars as requested by the handbox. The box can then synchronize its database to your sky. Most handboxes include a database containing the known positions of tens of thousands of bright reference stars, galaxies, and other targets. Better accuracy can easily be achieved using three-star alignment, rather than the far quicker one-star option. However, *goto* accuracy relies on good polar alignment as well. This latter process is often performed using *drift alignment*, which is described in a later chapter using real examples.

The whole mathematics and physics of computing star positions for driving your telescope relies on the computer calculating positions that may assume that your mount is horizontal and has an axis pointing to celestial north. Some degree of error in these parameters can be tolerated, but the more accurately they are set, the better should be the actual pointing when you activate the *goto* facility.

As discussed briefly, a beginner's telescope with a basic mount can be bought within the price range from perhaps $150 up to $400, depending on any offered additional facilities. The inclusion of a motor drive on one or both axes will have a marked effect on both usability and price. Of course, you can also consider buying second-hand. It is surely much better to spend a nominal amount now, rather than discover a loss of interest after spending several hundred dollars on something more advanced.

Mid-range Scopes

Even astronomers with high-end scopes usually keep a mid-range scope for wide field imaging or viewing. A larger scope cannot provide a wide-field view of large galaxies – unless a camera is mounted on the main tube pointing at the sky with a telephoto lens attached! Good advice before buying a new mid-range scope is clearly essential here. You need to know exactly what you plan to use the scope for before you commit serious money to buying one. There is a huge choice of makes, designs, quality of optics and construction, features of operation, sturdiness of mount, and ease of use. There are entire books devoted to the purchase of this range of telescopes.

When lunar eclipses are in the news, or perhaps a new, bright comet (and there were a number of these during 2005–2008), you may prefer to use a small refractor such as that illustrated in Fig. 1.5. This did not come with its own eyepiece, although one or two other accessories were included. With a good focusing mechanism built-in and a standard eyepiece fitting, it forms an excellent mid-range scope and is also suitable for autoguiding a main telescope.

One can easily acquire small telescopes from second-hand sales at a local astronomical society meeting. They may not be suitable for guiding a main telescope. These are best suited to use as nominal viewing scopes because focusing facilities may be limited and insubstantial – quite unsuitable for fitting a small camera. It is a good idea to buy a quality mount for such telescopes. Do your research. It is important to ensure that you specify the exact requirements that you have for your mount in order to avoid any misunderstanding with dealers. A mount such as the *HEQ5* with *SynScan* handbox – in its final form provides the computer facilities that many find essential – see Fig. 1.6.

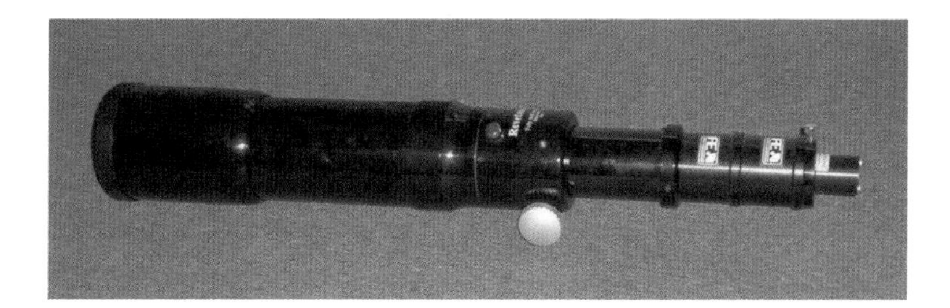

Fig. 1.5 An 80-mm refractor: ideal for wide-field views, when fitted to a suitable equatorial mount. Note the small CCD camera fitted to the eyepiece end of the telescope. In my case this is used for autoguiding (see later)

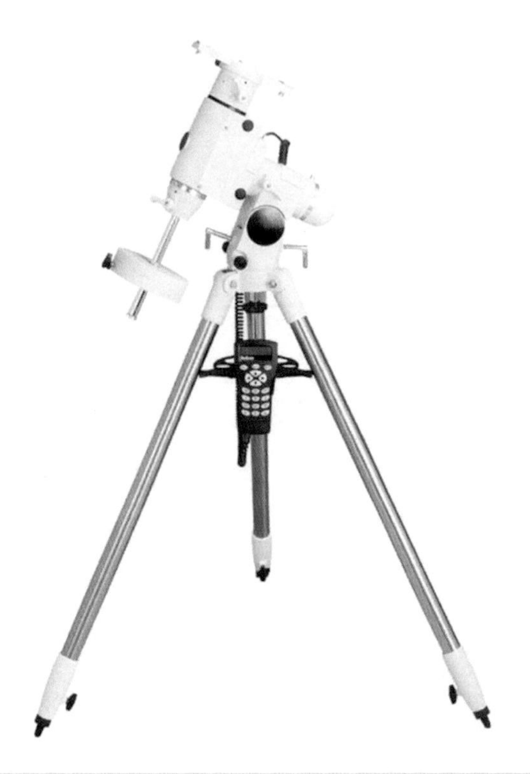

Fig. 1.6 The *HEQ5 Pro SynScan* mount with handbox controller

Wedges for Imaging

Although not essential for visual astronomy, a wedge is essential for imaging – see Chap. 3 on accessories. It allows the telescope to be tilted at an angle that, when equal to your latitude, allows the telescope to be driven on one axis only, instead of the two that would otherwise be required. Once carefully polar aligned, the mount should track reasonably well with any mid-range scope tubes fitted. Wedge adjustment facilities on these mounts are basic, as one would expect, but they should prove adequate for the job. The built-in software is an important consideration; this version can be updated by downloading firmware from the appropriate site. The handbox includes a large database of objects, but the main capability is the option to connect it to a computer for control by third-party programs. Most importantly, it can be controlled by *Maxim DL*, giving it huge versatility – see Chap. 5.

As with all mounts, a set of instructions is provided to aid the setting up of the tripod, the fitting of the balance weights, preliminary adjustment of the wedge to your latitude, and the operation of the handbox. Ensure that whatever mount you buy, the tripod legs are set on a firm surface. If soft grass is used, it is likely that the legs will sink into the ground to some extent, due to the combined weight of the scope plus equipment. It is possible to prepare a firm base even in soft ground. Dig holes where the tripod legs will rest, and fill the holes with stones, topped up with concrete. Small quantities of concrete are not expensive; a bag of ready-mix can be bought. Alternatively, lay a concrete base for the whole mount – an ideal solution. A firm base makes all the difference to ensuring good tracking following your extensive efforts to align the telescope.

The handbox included with this *HEQ5-SynScan* version has a fairly comprehensive menu for both initial setting up and optimizing, as well as routine observing. The unit also includes periodic error correction (see Chap. 6) and the option for autoguided observing. It is a common practice to mount a guide scope on the main scope and have the guider provide frequent position corrections to the mount.

The Importance of a Balancing System

With the mount should come a balancing system, normally one or more heavy weights that slide along the main axis – see also Fig. 1.6. Depending on the telescope design, the declination axis is usually balanced by suitable positioning of the main telescope fitted with either a camera or the eyepiece unit. Balancing is accomplished by cautiously releasing each locking control in turn, ensuring that there is no resulting violent swing. Adjusting the positions of the equipment can then be done. When properly balanced, the telescope and ancillary equipment should remain stable in virtually any position. It is acceptable to have a slight imbalance on the RA axis toward the east because the motors will drive westward at sidereal rate. A slight imbalance ensures continuous contact with the drive mechanism. Similarly, the declination axis can be left very slightly biased in either direction (up or down). Any residual significant moving tendency will probably strain the motors and should be avoided.

Equatorial Head

The equatorial head allows the fitting of a metal plate on which can be mounted the main scope as well as a smaller guide scope. The latter, as will be explained in detail later, can be used to monitor a guide star (which is not necessarily the target of the main scope) and to detect any tracking anomalies produced by the mount's imperfect gear train. These anomalies are registered on a separate guide camera, and corrections derived by the software are then fed to the mount's autoguider input. These small corrections help to keep the main target centered in the main scope, the process being designed to produce a sharp image. In Chap. 7, we explain the detailed process of calibrating and running a guide scope with the LX series telescopes.

Autoguiding Accessories for Imaging

Clearly, a telescope that offers the ability to autoguide is capable of producing significant astronomical images, rather than merely snapshots of a region of the sky. To do so, you need to buy a guide scope, such as that shown in Fig. 1.5, so anticipate adding this to the cost of producing high-quality images. Also allow for the extra cost of a guide scope CCD camera – see the unit fitted at the eyepiece end of Fig. 1.5 – though these are now available at competitive prices. Another important consideration at this stage is compatibility. If possible – and it usually is – try to buy the whole CCD camera assembly at the same time. Many CCD cameras are now fitted with a facility either already incorporating a guider or with a connector for a proprietary cable for connecting to a proprietary CCD guide camera.

Your Ideal Scope

As your experience of using a larger size of telescope increases, you will realize that several routine tasks could be made easier. Focusing, collimation, polar alignment, and date and time entry – these are a few of the tasks that must be checked before any real telescope success can be achieved, unless you are simply looking through the eyepiece! To have the telescope fulfilling its potential, almost everything must be set correctly. The more advanced telescopes incorporate facilities to make many of these tasks easy – or at least tolerable! A summary of these processes is given below, with more detailed operational descriptions in later chapters.

Time/Date Latitude/Longitude Entry

The telescope's computer has a built-in database that needs to be synchronized with the actual sky if you wish to use the facilities such as *goto*. With the telescope powered on, the handbox will request basic data entry. First enter your time/date and latitude/longitude where requested. Other parameters, as described in the telescope's manual, may be requested. One further entry – pointing it to at least one known star – is usually enough to fully synchronize your telescope.

Focusing

Focusers come in various forms, and their effectiveness will relate to the cost of your scope. There are simple mechanical types in which the eyepiece rack is moved backward and forward along its optic axis until good focus is achieved. Better telescopes may include electronic focusing; this is usually more controllable and reproducible and may provide a numerical display for reference. The position of best focus is likely to change during the night, due to thermal changes affecting the expansion of elements of the tube and optics. A change of eyepiece obviously requires a change of focus as well, because different focal lengths have to be accommodated.

Collimation

The best optical systems should maintain strict colinearity of the main mirror, the secondary mirror (or mirrors), and the eyepiece. A well-collimated telescope should show the best quality planetary and other images that are possible. Poor collimation leads to optical aberrations such as coma. With lower priced telescopes, such an adjustment facility is likely to be absent and probably unnecessary anyway. Telescopes having apertures above perhaps 15 cm benefit considerably from a facility to adjust the exact position of the main mirror; most Newtonians already have the necessary facilities to adjust the secondary mirror. Larger instruments, particularly Schmidt-Cassegrains, usually incorporate three tiny adjustment screws at the top of the scope on the back of the secondary (usually visible on the front of the corrector). By carefully adjusting these, good collimation can be achieved.

Polar Alignment

Although this is a function of the user accurately setting the mount's polar axis to point at the Earth's celestial pole, modern telescopes often have some means of helping achieve alignment. One type of facility is termed a *polarscope* and provides a see-through polar axis with graticule. This enables the user to visually roughly align the axis with the predicted position of the true north celestial pole (NCP). Note that Polaris – often used as a guide to the NCP – is about 1° off the true position. The graticule – or any good planetarium program – will show where the true celestial pole is in relation to Polaris.

Drive Accuracy

A telescope mount has to drive the actual telescope through the equivalent of one complete revolution of Earth every sidereal day. This length of day is the time taken for the stars to appear to move exactly once around Earth and is about 23 h, 56 min, and 4 s long. Most computerized telescopes include drive-selectable handboxes that can be set to deliver sidereal, solar, or lunar drive rate. This facility therefore helps considerably if using the telescope for viewing the Moon. Viewing the Sun, of course, requires exceptional care and attention to detail. There are special solar filters made by reputable companies for solar viewing, and on no account should any inferior product be used. *You cannot grow new eyes.*

Periodic Error Correction

Getting the best out of your telescope has never been easier, if still challenging! Suitable software can greatly improve the smooth running and tracking of those telescopes suitably equipped, whether amateur or professional. In a perfect world, the worm or wheel that drives the main shaft would accurately turn the telescope at exactly sidereal rate and in a perfectly uniform manner. No such luck! In the real world, the mechanical worm and gear teeth have imperfections – small tight or loose spots. The actual result of leaving the scope centered on a star and watching closely, preferably using a cross-wire eyepiece (a graticule), soon shows the combined error caused by variations in teeth accuracy and drive accuracy. The star will appear to drift forward a little, then backward, and repeat this motion seemingly randomly. In fact, the movements are likely to relate to errors on specific parts of the worm drive or other gearing within the gearbox. Consequently, these will probably be repeated every time that this part of the drive is reached. This type of drive error is called *periodic error*, and many higher priced mounts include a facility called *periodic error correction* to allow the user to modify it. At its simplest, this involves a visual monitoring of a suitable star. At its best, it means leaving a CCD camera recording the movements of a selected star and then analyzing the movements when perhaps three complete worm cycles have been recorded. By this means considerable improvement can be made to tracking accuracy.

Many of today's better telescopes have facilities to permit the correction of such imperfections, and this can enormously improve the resulting tracking – in some cases, right down to the limits imposed by the atmosphere! Much software is available for telescope performance enhancement and also for image analysis. This book includes sections on some of the significant software packages available for both purposes. There is no "one size fits all" program, although some packages are highly capable products.

Telescope Mounts

Any telescope mount costing up to about $2,000 is likely to include several, if not all of these facilities. Those nearer the lower end of the price range may include a fairly basic version of some facilities to complete these tasks – but at least they can be tackled. Those near the upper end of the price scale should provide reasonably comprehensive control. With most of these facilities present, you can anticipate a satisfying observing session that should not include a constant need to make repetitive adjustments to counteract poor motor driving and image drifting.

The Final Telescope Upgrade!

Some might say there is no such thing as a final telescope! Any scope will sooner or later suffer component failure – though hopefully not for at least a year and more. Repairs to smaller scopes often cause a reevaluation of longer term plans. This book makes the assumption that you want to move on from using a nominally sized instrument with basic facilities toward the advanced fully computerized scopes. At this point we can assume that your experience with medium aperture scopes, and a desire to move to a top-of-the-range model, has caused you to consider the LX200, and possibly even the LX400 series scopes. In many ways these scopes work in a similar manner, but there are also differences. The LX400 series has considerably more electronically controlled facilities than the corresponding LX200 telescope, and some design changes. The differences contribute to a more advanced telescope experience with the LX400 series. The consequence of this automation is the considerable increase in cost and a much heavier scope. The LX200 models have sold in far larger numbers.

Prices generally reflect the facilities offered by a telescope, and a top-of-the range scope can cost around $4,500 for a 30-cm (12-in.) model and well above that for larger models. UK potential purchasers have to buy these telescopes from the UK dealers, where prices are not directly related to currency. A typical UK price for the same telescope is £4,550.

Telescopes with apertures up to 40 cm (16 in.) can be bought; however, these are not normally for beginners. Such scopes are extremely heavy and large and are designed to be set up by universities and other institutions with budgets, land, and facilities available for planned observatories.

At the multithousand dollar level, there is a limited choice of manufacturer. Peruse the astronomy forums, and you will find references to *Meade*, *Celestron*, and manufacturers supplying even more costly telescopes. However, there is a limit to what even the keen amateur feels able to spend on what is essentially a hobby. This book features the LX200/LX400 Meade telescopes. With suitable accessories, a Meade LX200 (or 400) can be taken close to the limits of performance. The rest of this book is devoted to explaining how to get your proud possession to perform as you wish.

Chapter 2

Meade LX200GPS/ LX400 Series Telescopes

Getting Started

This author's own experiences with the LX200 series scope stretch over many years, starting with the purchase of a 10-in. (25-cm) LX200 (subsequently called the Classic) in 1992 – see Fig. 2.1. The picture shows the telescope mounted on an equatorial wedge inside his observatory. It was one of the early models. This chapter reviews the capabilities of recent medium-sized LX200-ACF telescopes (as the new models are now called), together with the enhancements offered by their companions in the LX400-ACF range (what was originally called the RCX400). The new letters ACF mean *advanced coma-free* optics. In later sections and in the next chapter, we also describe accessories thought to be essential.

The process of unpacking and assembling your telescope is fully covered in the telescope's manual, which should also include any recent modifications. The manual provides a description of many of the things that you need to know for successfully setting up the telescope, together with the necessary safety warnings. More advanced procedures are explained in this book.

What's in a Name?

The LX200 series scopes have endured various name changes over the years. The original LX200 was called the Classic and was followed by the LX200GPS when the GPS receiver was added. Improvements to the optical coatings and further enhancements were indicated by the additional letters SMT, followed by more

DOI 10.1007/978-1-4419-1775-1_2,

Fig. 2.1 A 12-in. (30-cm) LX200 Classic mounted on an equatorial wedge in Southampton back garden observatory with proud owner

name changes: /R and more recently ACF have indicated later modifications to the range. Here, we are adopting the following nomenclature:

LX200, LX200GPS, and LX400 as generic names
The LX200 is the Classic
The LX200GPS covers LX200GPS SMT – SCT/R/ACF
The LX400 covers RCX400 and LX400-ACF

During the early preparation of this book Meade announced changes to the RCX400 telescope range. Significantly, for legal reasons the name was changed to LX400-ACF (*Advanced Coma-Free*). It was decided that only the larger models – the 16 in. (40 cm) and 20 in. (50 cm) – would be manufactured in the future. However, there are still many smaller sized models (the 8, 10, 12, and 14-in.) in operation. Corresponding telescopes in the LX200 range were renamed LX200-ACF. The ACF models include the improved optical system originally designed for the LX400 scopes but remaining with f10 optics. In this chapter the terms LX200GPS and LX400 refer to the different scopes, both having ACF optics. The original Schmidt-Cassegrain telescope optics are found on thousands of older scopes in the LX200 series.

Meade LX200GPS and LX400

If you have been involved in serious computing, it is natural to want to apply that expertise to your telescope. Alternatively, if you have been using mid-range telescopes and found them a little limiting for your ambitions (as intimated in Chap. 1), you may have studied the market to find out what high-end equipment is available. The advent around 1992 of the computer-controlled Meade LX200 (Classic) series telescope and its sturdy mount was therefore a natural progression for many people. Recent versions have continued to add improved features, such as the focusing mechanism. A first experience of focusing with the LX200 will involve manually turning the focus knob and discovering that with such a design, the image being focused moved vigorously toward the edge of the field. This may be unexpected, so you have to master the art of focusing an SCT while periodically moving it slowly in right ascension to retain the image within the field of view. A regular manual winding of the focuser fully in and fully out helps to distribute the grease along the focuser shaft and consequently helps minimize *image shift* – as this movement is called. This still applies, as does doing a grease dispersal routine every few weeks. Another focusing hint: try to make your final focus twist a push so that the mirror moves away from the rear of the scope; this helps minimize focus backlash that might otherwise let the mirror sag. You also have the mirror lock facility to further improve this.

An electronic focuser add-on was a near essential accessory. The accumulating cost of accessories quickly leads one to realize that the owner does not merely pay for a telescope! Recent models in this series incorporate a microfocuser that largely eliminates the image shift problem.

Within a few months of buying the telescope, at the approach of autumn, you discovered the vulnerability of SCT corrector plates to condensation! Various devices were used to maintain a clear corrector plate. A dew shield is likely to be necessary, particularly in regions prone to high humidity. Light pollution may be severe, and a dew shield can shade the optics reasonably well in many cases. This also helps to limit the onset of dew, though only marginally. An electrically heated band to wrap around the front end of the tube prove – another essential purchase – the *Kendrick* system (see Chap. 3) being a good example.

LX400 scopes are considerably more advanced, having an additional set of focus motors in the optical tube assembly (OTA) – see Fig. 2.2 – apart from numerous other useful features. The focus feature provides precise handbox adjustments to both focus and collimation, insuring the ease of obtaining the highest quality imagery. The corrector plate does not have collimation screws because of this electronic control. Because of the extra motors these telescopes are much heavier than their predecessors (the LX200 series). This chapter takes you from opening the box, insuring adequate power supplies, understanding the importance of initial balance and alignment, through to Meade's *Smart Mount Technology (SMT)* feature (explained later). Actual detailed descriptions of balance, polar alignment, and other processes are discussed in later, separate chapters.

What You Get in the Box

Because this book deals in detail with the advanced Meade LX200GPS (and includes personal experiences with the top-of-the-range LX400 model as well as

Fig. 2.2 The front end of the 30-cm LX400 showing the corrector plate. Focusing is achieved by motors driving the corrector plate within the OTA

many years with earlier LX200 models), let us assume that you have taken delivery of one of these fine scopes! You have (hopefully) already studied the online manual that can be downloaded from Meade's Web site in advance of delivery (http://www.meade.com/manuals), and so know exactly what you have bought – and even how much it weighs! The box packaging includes a printed manual, and there are other boxes that house the various optical components such as the large, heavy eyepiece and the right-angled diagonal mirror. Although the LX400 comes with software on a CD (the essential Meade LX400 USB driver and *AutoStar Suite* software), the LX-series scopes currently exclude this. However, this *AutoStar Suite* can be downloaded directly from Meade's site. The LX400 includes a USB connection cable. The finder and mounting components are carefully packaged as well. A tip: take some pictures of the layout of the boxes and the positioning of their main components (see Fig. 2.3). If you ever have to re-box the telescope for sending back to Meade or your dealer for repair, you could find these pictures invaluable!

The 12-in. LX200 (Classic) is at the limit that most people can manage to carry, keeping in mind that it is often used on an equatorial wedge and therefore has to be lifted higher and at an awkward angle. You may have some heavyweight friends who can do this job, available from the local astronomical society, but be careful here. Do not risk any of your friends or neighbors damaging their backs on your behalf. Hire someone if you have to. Some people are trained in the art of lifting heavy objects.

The telescope manual describes in detail how to assemble the scope and orientate it prior to mounting. Ensure that you understand exactly what has to be done with the various accessories before construction. If you assemble the scope in an observatory, you can do this without major concern about the weather. My first 10-in.

Fig. 2.3 LX400ACF accessories box as first opened. The package includes the handbox, finderscope, right-angled prism, cables, and assorted adapters. A picture can be very helpful should you need to repack for transport

(25-cm) LX200 was for outside, uncovered use, and so the weather was a major consideration. An interest in weather satellite monitoring provides an excellent method of seeing live weather data that can help personal forecasting (see Chap. 10).

LX200GPS and LX400 Connector Panels

On the standard LX200GPS telescopes there are no USB connectors; all commands are sent to the scope via the RS232 ports, as shown connected in Fig. 2.4. The front connector panel of the LX400 telescope is different – see Fig. 2.5 – offering a USB connector. The rear of the LX400 optical tube assembly – see Fig. 2.6 – is also unlike that of the LX200GPS series scopes. There are three USB connectors (seen in Fig. 2.6) that access the built-in hub; these offer direct connection for the CCD camera that you will surely wish to use. One caveat here: long experience indicates that in colder temperatures (possibly below 5°C) the built-in hub becomes unstable. Although for most sessions you can reliably use this connection to save the long USB run from camera to computer, under low-temperature conditions error messages announcing the loss of connection can interrupt an imaging session. If you are planning on leaving the telescope under full automation overnight, as described in later chapters discussing advanced software, consider the alternatives. Many astronomers successfully use an external powered USB hub. Another wiring option is that of a direct cable run to the computer.

Fig. 2.4 Front panel of the LX200GPS telescope. The electronic focuser connection is to the right of the power connection; other connectors (*left to right*) are for the handbox, serial connectors (RS232), and the autoguider port

Fig. 2.5 Front panel of the LX400 telescope. Connectors left to right are for USB, auxiliary connection (*see text*), main handbox, and a serial connector

Fig. 2.6 Rear view of the LX400 showing handles, cooling fan, central visual back seal, and OTA connector panel

For visual use, a reticle connector is available to plug in a reticle eyepiece lead to provide power for the reticle's built-in cross wires. A second handbox connector is available on the LX400 OTA, though it is hard to imagine any circumstances in which it could be useful. The RS232 and autoguider inputs are ideally located; use the autoguider input for direct input of correction signals from whichever system you are currently using for autoguiding (see Chap. 7). The smart accessory connector is a feature reserved for the large *MaxMount* (40-cm and 50-cm models).

The Equatorial Wedge

No wedge is included with the telescope unless you specifically order one. Let us review the decision about whether you are likely to want to use a wedge in the first place. If you are a purely visual observer, the good news is that it is not really necessary to install a wedge. When used in the basic alt-azimuth mode and fitted with the supplied eyepiece, the telescope is well balanced in both altitude and azimuth. The only consideration for the visual observer is that objects will not normally be orientated north to south in the field of view. As the telescope slowly rotates on its axes, the object and field appear to slowly rotate within the eyepiece view. This is not really a concern, though it does result in planetary satellites – such as those of Jupiter – not usually being viewed in the horizontal plane. The various special functions of the telescope, including focusing, collimation, *goto,* etc., all operate exactly as they should, assuming that the telescope is set up for alt-azimuth mode!

Additionally, there is no need to do any PE (periodic error) measurement or correction. Even a scope with poor PE (see Chap. 6 for a full explanation of this) will show hardly any movement as far as the eye is concerned! However, if your plan has been to stick with purely visual observing you would not really need the advanced features of these high-end scopes. Other, cheaper models having an identical aperture would surely be more than adequate.

For those wanting to specialize in astroimaging, there are two possibilities. An equatorial wedge – see Fig. 3.2 in Chap. 3 and Fig. 4.1 in Chap. 4 – is an excellent, almost essential device for converting your dual-purpose telescope into a fully functioning imaging scope. Before that desirable status can be achieved, however, you must go through the process of balancing and polar aligning your telescope. Also covered in detail later is the initial setting up of the equatorial wedge.

Field Derotator

The other option, particularly for those living near equatorial regions (where low latitudes make it exceptionally impractical to use an equatorial wedge), is a field derotator. This device integrates with your scope's *AutoStar II* computer to enable imaging with the scope in alt-azimuth mode. The scope does not therefore require an equatorial wedge when used this way. The *AutoStar II* calculates the amount of field rotation during an alt-azimuth imaging session based on the region of sky being imaged. It then rotates the field derotator and attached camera in the opposite direction in order to precisely compensate for the field rotation.

The advantages of a field derotator include less additional weight (than a wedge) for field use and less time needed for polar alignment. You can also image any part of the sky with the field derotator. However, the equipment stack (flip mirror system and camera) at the rear of your scope might be too long to pass through the drive base of your scope when used near the zenith. The backlash experienced when crossing the meridian can also be a problem. Another reported problem is the difficulty experienced when using PEC – the periodic error correction built-in feature for use in polar mounted scopes – in alt-azimuth mode. Although PEC is (technically) available, many consider it to be ineffective in this mode. Telescopes with poor periodic error may not be able to provide acceptable long-exposure images. Chapter 6 looks at the measurement and correction of periodic error.

Before deciding on whether to buy a wedge or a derotator, you have to consider the advantages and disadvantages to you and then decide which way to go. A large majority of astronomers seem to opt to install an equatorial wedge.

For our purposes, let us assume that you have followed the manual and have mounted the entire optical tube and fork assembly directly onto the tripod. If you wish to mount the telescope directly onto a wedge, you should first read the section in Chap. 4 devoted to setting up your wedge. Either way, let the excitement commence!

Power Considerations

The next major consideration following the heavy work of mounting the scope is powering it up. Depending on your exact model – and therefore its power requirements – it is essential that you obtain and use an adequate power supply. Many of the Meade scopes come with a battery compartment, and the manuals imply that the scope can be run using a number of battery cells. Consider accepting the wisdom of engineering, other experts, and long-time users on the LX forums, and – instead of internal batteries – either use an adequate external battery (which also makes the scope portable as well as providing a clean supply) or use a fully regulated main power supply of adequate power, such as a 10 A power supply (PSU). Beware of recommendations for power on a telescope supplier's Web site. Without enough power there could be immediate problems with the focuser, memory, and even simple driving tasks. The LX400 in particular requires the availability of heavy power and a minimum of 10 A. Do ensure that any main power supply is adequately clean, electrically speaking. A noisy supply can adversely affect the scope.

Caution: It is essential to insure that your supply has reverse polarity checking as well as suitable fusing. It appears that the scopes have no built-in protection against these problems. For the very first use of an LX scope, it seems eminently sensible to set it up for visual use using the supplied optics. This has the benefit of insuring that the scope is already nominally balanced. In this configuration it is perfectly safe to test virtually all of the scope's features. The manual shows where the handbox is connected and mentions the availability – in the case of the LX400 model – of a second connection socket on the back of the OTA itself (see Fig. 2.6). Special care must be taken at the time of first power up, as explained in the manual. At this time, on selecting *autoalign*, the scope – whether LX200GPS or LX400 – is programmed to swing around seemingly at random, although this is, of course, following a preset sequence. On that first power on, you might watch your LX400 drive to various positions using its activated sensors to identify north and horizontal. When it finally finishes this initial alignment, it "knows" where it is and the exact time (using the built-in GPS receiver) and therefore calculates that it is no longer in its place of manufacture, Irvine, CA. The manual describes the process of editing in your own locality name for future reference, and this is straightforward.

Needless to say, this initial alignment process might be different on your scope. The development of increasingly sophisticated software and firmware may result in a change of initial alignment operations. It is essential that the instructions in the manual are read carefully before the first power on. Later, when the *AutoStar II* (telescope) software is investigated, you may find that there are discrepancies between the contents of the manual and the actual command sequences provided! These differences are occasionally discussed on the forums.

Sun Warning!

Much of your initial testing can be carried out during daytime. In that case, you should leave the end (corrector plate) cover on because it is *imperative* that both the telescope OTA and the finder have their covers securely fitted. Under no circumstances should the Sun be allowed to enter either optical system. The intense heat and light could severely damage the scope and might have other unexpected consequences. Following the power-on test, you can at least confirm that nothing untoward happens. You could also set up your computer to connect with the scope during this initial test session. However, you are unlikely to be able to synchronize the scope with any astronomical object, and it is not important to be able to do so at this stage.

GPS Receivers

Once the telescope has completed the initial self-alignment process, you should also see the self-check of the position of the index mark used to synchronize the scope's PEC table (which will be explained later). During the process that follows power on, the telescope mount drives approximately 24° westward, as described in the manual. A worthwhile action here is to use the handbox *setup* option to switch the GPS receiver to receive *GPS at start-up*. It will then, in future, cause the scope to seek GPS satellite data after power up, insuring that the scope "knows" the precise date and time. This information is not stored on the scope. If the PEC table has been activated (see Chap. 6) *and* the scope previously *parked*, it will not need to do this westward drive each time it is powered up. If you are in a poor reception region for the GPS satellites, or if you prefer, you can disable the GPS receiver and manually input the time yourself at every power on.

One possibility here is that the GPS receiver could fail to achieve a fix during this first time. You can try loosening the RA clutch and moving the tube to see whether the GPS receiver can get a better fix. If you are not prompted for your site, then the scope may remain set for Irvine. In that case you should manually *edit* your custom site using your – at least approximate – coordinates. It is also essential to *edit* – if not performed by the GPS fix – your time zone. Be aware that both LX200GPS and LX400 use negative values for the difference from GMT for longitudes west of Greenwich, the reverse of the convention used by the Classic LX200!

To power off the scope and keep most of the previously adjusted settings, select *utilities, park scope* (perhaps not the obvious menu item to select); this simple act will cause the scope to drive to a known *park* position and (in the case of the LX400) will save the last *focus preset* and other data for availability at power on. If you had changed the focus before *parking* (this is quite likely) then problems can occur with reset data. You can read a full description of the *park* process in the manual and can subsequently *define* your own *park* position. Try using the standard due south on the celestial equator *park* position.

Once your scope is fully operational and has been correctly aligned and synchronized, you should find it very easy to locate several of the planets and even bright stars during the daytime – visually. Consider doing this when Mercury, Venus, Mars, or Jupiter are favorably placed. However, there are caveats that go with such special observations, so this type of viewing cannot be recommended at this stage; wait until you have set up your scope properly.

First Visual Observing Sessions: Finder Adjustment, Focusing, Collimation

It is clearly important to be able to focus your telescope easily. Follow the manual's recommended sequence of operations for your first telescope session, including backlash adjustment. This provides an opportunity to identify a specific landmark for your reference point. A preliminary focusing of the telescope can then be done at this stage while locating your landmark feature. After the backlash measurements and adjustments are completed, you can use the same marker to align the finder – see Fig. 2.7 – precisely. Aligning the viewfinder in the daytime during the first power on session is an excellent strategy.

Although this can also be done at night, the first daytime operation should allow many settings to be achieved before the first evening or night time observing session.

Fig. 2.7 Meade LX200GPS/LX400 Finder mounted on an LX400. The guide scope can be seen above it. It is important to align the finder on a bright target object visible in the main scope. This allows easy guidescope use

Terrestrial focusing will not be precise because the stars are at infinity, but it will be sufficient to enable a later precise focusing when you locate your first star field during the evening. The manual explains how to adjust the finder; four of the six alignment screws can be seen in Fig. 2.7. With the main scope focused and centered (preferably using a reticle eyepiece) on a bright, easily identified star, the finder can be adjusted using these screws until it is centered on the same bright star.

Focusing

The LX200 Classic came with a manual focuser that had idiosyncrasies of its own, as mentioned before. The design of the telescope includes a focuser that manually moves the mirror within a carrier tube backward and forward. This carrier tube is concentric with the telescope's baffle tube (a tube carefully sized and positioned to minimize the possibility of extraneous off-axis light reaching the mirror). The mirror carrier slides along it and experiences some rocking due to mechanical looseness within the system. Focuser adjustments often cause the image to slide across the field of view during the focus process. After considerable practice and manual adjustment, this effect can be minimized by fully winding the manual focus in both directions a couple of times every few weeks in order to better redistribute the grease.

More recent models in the LX200GPS series have a mirror lock and zero image shift microfocusers ushering in a new era of easier operation. As explained in the manual, a first rough (approximate) focus is obtained using the focus knob at the base of the Optical Tube Assembly (OTA). Remember the earlier hint to have the final twist push the mirror to help minimize focuser backlash. The mirror lock, positioned above the coarse focuser, is then rotated towards *lock* to a firm feel to lock the mirror in place. This should eliminate any later mirror flop, a problem observed on some occasions with earlier models when the image crossed the meridian. The microfocuser is used to obtain fine focusing with the advantage of no associated lateral image shift. Before making any further adjustments to the main focuser, it is important to release the mirror lock.

The LX400 model includes a feature incorporating electronic focusing with memory. To illustrate, you can focus the telescope with your current optical setup – whether for visual use with a specific eyepiece or imaging use with a specific focal reducer. This position can be stored as a labeled focus position (the display shows its position in millimeters) and subsequently recalled. You can therefore have settings for the basic f8 optical train using the included 24-mm eyepiece and another for your f6 CCD camera focus point. After swapping the optics, the desired focus position can be called up, activated, and the motors will drive to the recorded position. This is an excellent feature that can be used for changing focus from a night time imaging session to a daytime visual planetary viewing! Several potential storage positions are available, therefore offering multiple-stored optical configurations. Although the separate focus motors installed within the tube add greatly to the weight of the scope, they are extremely useful!

Experience shows that the firmware does require a little focus reminder before the end of a session. This can be performed as follows: prior to the *park* command, in order to properly retain the LX400 focus points, select a *focus preset* (your current application), then the command *park*. There is evidence that *AutoStar II* does not otherwise properly store the focus position, requiring you to manually resynchronize the focus to one of the known positions. It appears that *AutoStar II* only retains which *preset* you are at (one through nine) between sessions, and the *mm-from-zero* that each *preset* represents. At the next power up, it does not apparently remember where it was relative to zero, but if it knows it is at *preset #5* then it can work out where the other presets are.

Hartmann or Bahtinov Mask

A Hartmann Mask is very useful for helping to easily achieve accurate focus. The mask is essentially a cover for the end of the telescope tube and has three (sometimes two) large holes symmetrically placed. This effectively creates three smaller telescopes and therefore three separate images. The mask is placed in position for focusing, at which time the three out-of-focus images should become visible (see Chap. 5, Fig. 5.4), forming a triangle – assuming that focus is somewhere near. For easiest focus, a bright star should be near the center of your image. As you adjust the rough focus, the three images will converge. Using the electronic microfocuser, a near perfect focus should be obtainable. Do remember to remove the mask before continuing!

Some posts to the LX200GPS forums have referred to the Bahtinov mask (see Chap. 3, Fig. 3.10) for assistance with focusing. This mask covers the front of the OTA and has a series of slots cut into its material, producing a complex pattern. For perfect focus the central line in the star pattern should be exactly between the two diagonal lines. See the "Accessories" section of this book for more details.

Changing Speed

Right at the start of the tests with your new scope, it is strongly recommended that you set the scope's maximum *goto* speed to about 1.5° (many use 3°) per second. The scope's default speed is at maximum, so any use of the *goto* option sets your scope driving at top speed. Nothing – apart from a badly balanced scope – is more likely to quickly wear out your gears! Setting a new maximum speed is just one parameter that you can investigate during the first few sessions; during this time you can gain familiarity with the handbox controls. It might be useful to print out a copy of the entire command listing as provided in the manual. This list is retained in the observatory. Be aware that some recent modifications to the menu structure are not reflected in the manual.

LX400 (and LX200GPS with Care) Handbox Removal

Another useful option that may not come to light for a while is the removal of the LX400's handbox during a session. The LX400 handbox can be removed once the scope is powered up and can be (hot) reconnected later for power down. This is mostly of use during computer-driven (remote) operations when you are sitting in the house and controlling the telescope during a home network session. This is many people's normal operational mode. During very cold sessions the handbox is best removed from the scope and kept indoors (or at least protected in an insulated bag) in relative warmth. Many users feel that the inherent design of the handbox does not optimize its operation in very cold conditions. Often keys will not operate correctly, or the display will become distorted in such weather. When the liquid crystal display (LCD) cools down, the motion of the suspended crystals will slow in response to the changing electric fields. The formation of dew on the switch contact pads inside the handbox effectively short-circuits (simulating presses on) many of the keys at once. The handbox can therefore be confused by the false signals and probably will ignore genuine ones. Having a suitable cover for the handbox under such conditions is an important thing.

The handbox can be reconnected at the end of the session when you are ready to park the instrument. Note that you may need to hold the ? key to achieve a resynchronization of the *AutoStar II*. The LX200GPS handbox is reportedly also hot pluggable, followed by the key press to update the display. Many software applications, including Meade's *AutoStar Suite,* will control the telescope without the presence of the handbox, including parking.

Collimation

Your Meade LX series scope is a precision, advanced optical system and, as such, it deserves a precise alignment of its optics. The process of ensuring centralization of the mirror with the secondary in relation to the field of view is known as collimation. When the scope leaves the manufacturer the optics are accurately collimated; journeying across the continent(s) is likely to have some effect – however minor – on this alignment. The LX200GPS and LX400 series scopes have features that allow for the precise alignment of these optics, and the manuals give full instructions on performing this process. The LX200GPS has three adjustment screws set in the front end of the telescope. The LX400 has built-in focus motors that can be controlled electronically for adjustment by the handbox. In each case, the relevant manual describes the process, and both processes – as described here – can be done within perhaps an hour. Use a high elevation bright star (thereby minimizing the effects of atmospheric turbulence), making the adjustments after defocusing (see Fig. 2.8 and next section). Collimation procedures are best performed after any significant optical change, such as after changing from

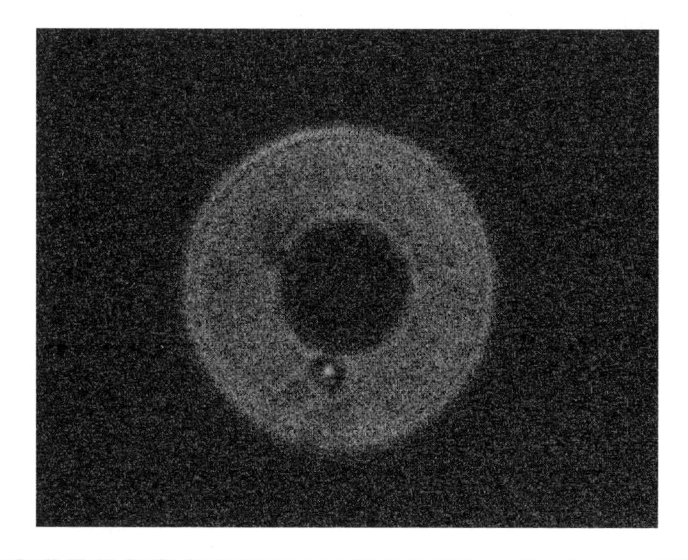

Fig. 2.8 An out-of-focus image of a star used during manual collimation of the scope

f10 to f6.3 or whatever comparable changes you decide to make for specific imaging projects. Such readjustments may be minimal.

To proceed with collimation, insure that the scope is temperature stable and then, as mentioned previously, select a fairly bright, high elevation star. The process involves defocusing the star, as shown in Fig. 2.8, and carefully examining the image. The central dark area is actually the shadow of the secondary mirror; it may or may not be centered in relation to the unfocused star. If it is off center, the optics require collimation adjustment. By carefully placing one finger on one of the three set screws on the black plastic secondary mirror support, you can identify which screw needs to be adjusted as follows: slowly move your finger until it is seen to be over the thinnest part of the concentric circles. Your finger will be either over one screw or between two screws, pointing at the third. The process of adjustment will cause the defocused star to cross the field of view, so to prepare for this, move the field at the slowest drive rate in the same direction as the darkest offset. The identified screw is then turned with extreme caution and by only a very small amount. By repeatedly checking the result of this adjustment, you should be able to obtain concentricity. The accuracy of this process depends on your judgment; however, in a later chapter on advanced software, we will explain about the use of software that can help quantify the collimation and achieve a surprisingly good result.

The manual provides a useful description of the process. After completing a basic collimation using the 26-mm (supplied) eyepiece, if you wish to further optimize the collimation you can use a 6-mm orthoscopic eyepiece for precise, sensitive adjustment. Accurate collimation results in the best quality views by enabling the best possible focus obtainable from your instrument. In turn, this results in the scope delivering the deepest views possible under the current circumstances. Collimation should normally take no more than about 20 min or so when done this way (manually) or possibly even less using the LX400 with its electronic collimation feature. It should only need to be done following optical configuration changes.

LX400 Collimation

Uniquely, the LX400 is collimated electronically using the handbox in collimation mode. The instructions are similar; only the method has changed. Key presses substitute for delicate screw turns. Additionally, there is a factory default setting available, so if you wish to retrieve the original factory-defined collimation, you can activate this option. When activated, the focus motors run at maximum speed until they reach the end of their travel; then they run back at maximum speed until getting close to the right encoder reading. They slow down until they are in the default collimation. Hearing your focus motors run at high speed is an unnerving experience; be prepared.

When you have achieved your best collimation setting you can save it (only in the LX400 scope) using the *collimation, save as default* option. On startup you could then do a *set to factory collimation*, thereby ensuring that the focus encoders record exactly where the corrector is. It is useful to realize that if your LX400 focus motors inadvertently enter runaway mode – a highly undesirable state – as well as hitting the mode button as quickly as possible to try to terminate this, you can subsequently activate the factory collimation default if your motors hit the stops. Focus motor runaway can occur on occasions when it is not expected – such as when using a focus preset – so always be prepared to hit the mode button; this will normally terminate the current action.

AutoStar Suite Operations

There is much useful software available for the LX200GPS series scopes, and it can be downloaded from Meade's Web site (see below). The *AutoStar Suite* comprises a set of features, all of which can be accessed within the program: a sky map (planetarium section), telescope control, the *AutoStar* image capture and processing facility, and the *AutoStar Update* option (see Fig. 2.9). The latter option can also be started independently if required. During your early sessions with the scope it is recommended that you familiarize yourself with some aspects of this software, particularly the planetarium part and mount control (see next section).

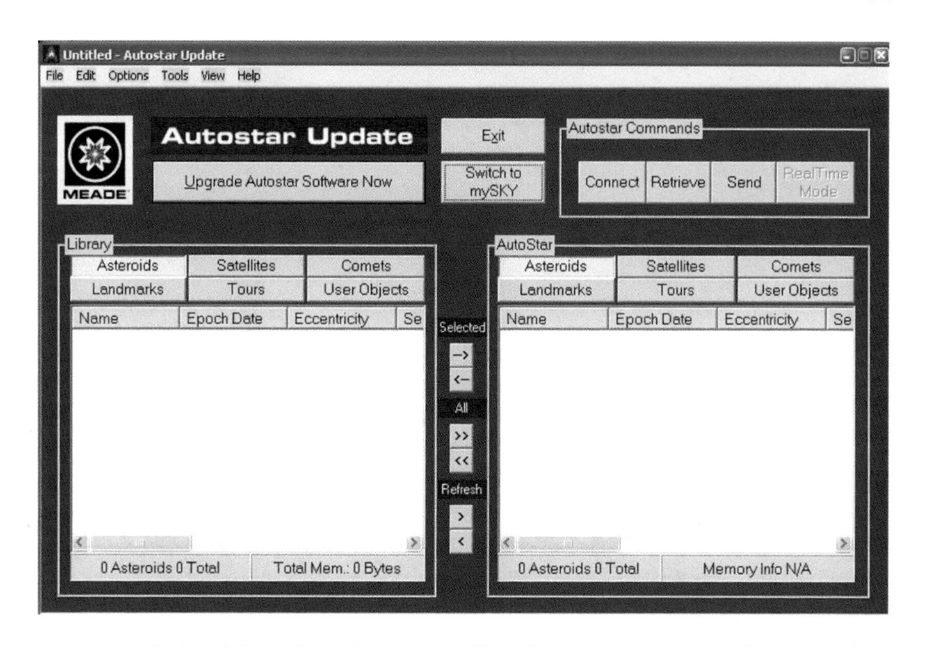

Fig. 2.9 The *AutoStar Update* program. Used for loading tours and firmware updates, etc. into *AutoStar II*

The similar-sounding *AutoStar II* software controls the telescope via the handbox. At a later stage, you may want to look at the mount modeling mode that Meade calls Smart Mount Technology (SMT) – a powerful option within the *AutoStar II* software. This is activated via the handbox, in which you select and identify a number of bright stars and center them in the field of view. The scope's software records these positional results and produces a mathematical model of the sky in relation to the mount's recorded position. When the model is complete and activated, pointing accuracy – as used by the *goto* facility – should be improved. The word "should" is significant here! As will be seen in a later chapter, bugs in the SMT can badly affect localized pointing. Although the scope's own printed manual does not include much material about the *AutoStar Suite* itself, it does briefly explain its operation. The LX400 is provided with a CD that includes these programs, along with other software and detailed descriptions.

The *AutoStar Suite* includes a help section that describes most if not all options. However, it is not difficult to master the application of the program. It is recommended that you use the *AutoStar Suite* and its planetarium component to confirm that your scope will connect to the computer using the supplied cable (where appropriate), and that it can be driven around the sky via the planetarium component. Should there be any queries about telescope control, use this software to check that the entire system is working normally. If you cannot control your scope using the *AutoStarSuite*, you are unlikely to be able to use it with a third-party program.

One word of caution: at this early stage, do not drive your telescope around without actually being there – for example, using a remote control facility or via a network. The scope has no way of knowing whether a cable is catching on something, getting taut, or whether something unexpected is in the way.

AutoStar Suite Updates

From time to time Meade has issued updates to the LX200GPS and LX400 series firmware and software. All such upgrades can be downloaded from the Meade site and then installed using the *AutoStar Update* software. Later versions of the LX200GPS telescope utilize improved firmware by using flash memory. Although these updates can be spotted if you have a regular look at Meade's Web site, it is far easier to continue to monitor the relevant forums where announcements are made about upgraded software. However, you should master the use of whichever firmware version comes with your telescope, rather than immediately try to upgrade it when a new version is released. Updating the *AutoStar* or *AutoStar II* is not a task to be performed without familiarization. We look at this upgrade process later in this chapter.

Meade's download site: http://www.meade.com/support/downloads.html

Connection to AutoStar Suite

Assuming that you have already installed the *AutoStar Suite* and followed the advice in the manual to install the telescope drivers, connection from the telescope to the computer running the *AutoStar* program should be straightforward. The LX200GPS series telescopes use a serial connection (see Fig. 2.4) for commanding telescope movements; the LX400 series can use either a USB connection (using the telescope's USB port) or a serial connection. Operations are therefore slightly different for each connection sequence. A useful tip to follow is that whenever you are connecting any cable between the computer and telescope you should call up the *operating system's device manager.* This instantly displays the actual com port used by the scope. This information may be required when configuring software for telescope control. Note that these details do not apply to Mac OS X users.

Your computer – particularly a modern laptop – is unlikely to have a serial port, so a USB-to-serial adapter may be essential. Use one to control your *Adaptive Optics* unit (see Chap. 9); you may need one for telescope control. Success has been reported using the Keyspan USA-19HS USB-to-serial adapter, and the associated device driver retains its COM port number when moved to different USB ports.

Depending on which scope you have, connect the supplied cable to your computer's port and the other end to the special connector on the telescope's main panel. With the *AutoStar* planetarium suite running, select *telescope, protocol* where you can ensure that the *none* option (if using USB) is ticked – or other number

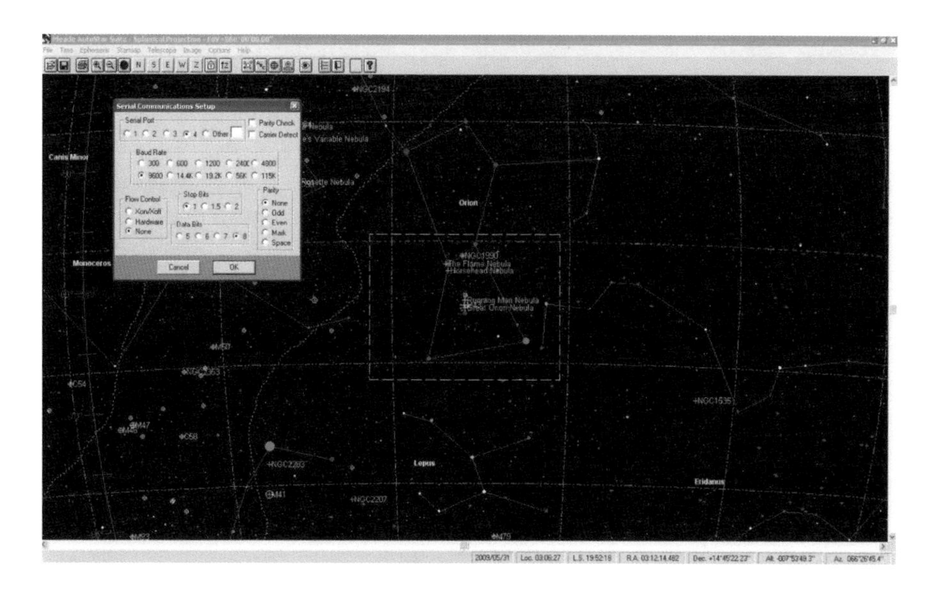

Fig. 2.10 *AutoStar*-COM port selection. Ensure that the correct COM port number is entered

as shown on the device manager display. Control can also be performed using a modem or network. The telescope end of the cable will see a com port because there is a built-in USB to serial adapter within the LX400 panel. If you now call up *device manager* (found on the computer's control panel option) you should see the active com port listed as *Prolific*. This will be the port number that you select for communications. The *AutoStar Suite* software offers radio buttons for COM ports 1–4, but you can enter a different number after you identify the correct COM port. Before you plug in the USB connectors into the computer, always activate *device manager* to see which com ports have been selected by Windows. Usually it is the same numbers – but not always!

Select *telescope, communications, com port setup* – see Fig. 2.10 – and select your *transfer speed*. If you are upgrading the handbox firmware, you can select a much higher speed for the transfer process. With the transfer speed at 9,600 baud, the update can take over an hour! Advice on the LX400 forum was to change the speed to 56 k baud. The upgrade should take only minutes. For routine telescope control it is essential to ensure that the speed is set to 9,600 baud in order to ensure that control commands are correctly interpreted.

Telescope Control

With all settings correct and the cable connected, you should be able to use the *AutoStar* program to control the telescope, as described in the online (pdf) manual. Targets can be selected and the telescope commanded to move to them. Similarly, the star map can be told to move to the telescope's reported coordinates. This process can be tested during the day to ensure that the telescope moves under control. If there are any obvious problems, they can be fixed during the day, rather than spending precious night time hours faultfinding under conditions of semidarkness.

Menu Tree

Use with caution! With Meade's *AutoStar Suite* connected you can start the remote handbox (*Telescope/Protocol/Remote Handbox*); you will see that it forms a complete tree view (with [+] to expand the branches) of your entire menu in the left frame of that application. You can now see where all the options are, one or two of which may not be as indicated in the manual. Do be careful before using some of the less frequently used options; some entries may not be valid all the time. One user cites the case of the *Ambient Temperature*, a display available on the handbox that only appears in the menu system after 10 min have elapsed from the time of the GPS Fix. This allows the thermal sensing hardware to stabilize and to ignore the nearby warm GPS receiver. Although always accessible from the remote handbox tree it will simply report the wrong value.

When your first night time session arrives, using visual mode (with the supplied eyepiece) you have the opportunity to master several features, including the focuser. If focus was last done using the terrestrial target during backlash adjustment, it should already be close to precise focus. Drive the scope to a bright star and center it in the finder's field of view that was previously aligned during the day. The LX200GPS has the mirror lock knob (previously discussed) that can be moderately tightened to minimize the mirror's tendency to move during some telescope movements. Before adjusting the coarse focus this lock should be released. Adjust it for best rough focus. The microfocuser can be adjusted to one of four speeds and then used to provide fine focus. This position can be stored on the LX400; note that the LX400 has five focus speeds. The correct procedure for storing positions is as follows: use *define*, and after confirming the label, hold the *enter* key for a couple of seconds to produce the beep. With the bright star centered in the eyepiece view, make any adjustments needed to ensure that the finder is also centered. This will ensure that when any required target is centered in the viewfinder scope, it will be well within the field of view of the main scope.

When using a CCD camera and the associated software as described later in this book, you can monitor the real-time analysis of a star's image to obtain best focus. This removes the need to scrutinize the visual appearance of the star through an eyepiece – a tiring experience. An electronic focuser was one of the early accessories of the LX200 Classic.

Note that the *AutoStar* software does not seem to fully control the LX400. Some tests indicate that you can make only one move of the focuser either in or out; proper focusing appears to be impossible.

Updating ASU and Firmware

As referred to earlier, updating the software/firmware should not be undertaken until you are experienced at using the telescope and software; there are pitfalls for the novice user, so be aware. A further word of caution: run your computer from the mains – not its battery! Should your battery fail during the update, you will probably need to do an emergency reload (see later notes on *Restoring a Corrupted Handbox*).

When you are finally ready, download the latest *AutoStar Update* (*ASU*) application from Meade's Web site (assuming that you do not already have it installed). The *ASU* software features a connection between the computer and the telescope, providing a communications link via the cable through which firmware updates can be sent. The program offers access to the handbox contents such as asteroids, landmarks, and tours. These can be retrieved as well as modified, making it a fine program for producing tours of the sky for star parties, as well as for the serious use of updating the firmware.

Download the latest firmware version for installation in the scope's computer – if not already done. The firmware provides calculations for pointing as well as other features. Install the *ASU* application on your computer. The downloaded *ASU* file can be run and the instructions on the screen then followed; accept the default settings. The firmware can be unzipped and the final ROM file placed in the *Ephemerides* folder under the *ASU* application folder. All the instructions are provided.

The sequence can be summarized as follows:

1. The telescope and computer can be powered up. If you have a screensaver running, disable this to insure that there are no sudden breaks in operation.
2. Connect the standard LX200GPS/LX400 interface cable between your computer's serial port and the leftmost RS232 port on the telescope's control panel (next to the HBX port).
3. Run the *ASU* program. Check that the proper COM port has been selected.
4. Select *Connect*; then set the maximum baud rate to 57,600.

SMT recordings (see later explanation), PEC data, and all saved user objects should survive a firmware update. It is, however, likely that you could lose *Site*(s) data, *Training* and *Calibrate Sensors* results (applicable if your scope is alt-azimuth mounted). If you have data that you wish to save, you can retrieve and save it as follows:

Retrieve hand controller data by pressing *Retrieve* under *AutoStar* Commands. Save the data via *File/Save Hbx Data*.

Although an upgrade itself should not corrupt PEC data, this data may be corrupted by the *AutoStar Save Hbx Data* feature, so this should be excluded from the save process.

Upgrading Firmware

Continuing with the upgrade process: Select *Upgrade AutoStar Software Now.*
In the dialog box select *Local* and insure that the latest version of the firmware previously placed in the *Ephemerides* folder (as described above) shows under *Select* from versions on the hard drive.
Leave *Erase User Banks* unchecked. Click *OK*.
Respond *OK* when the prompt displays.
Respond *OK* when the *Clear User Data* box opens.
The message confirms all is well, so press *OK*.

The lower left corner of the *ASU* application now displays the progress of the update. The hand controller display shows: *Downloading. Do not turn off.*

If you previously set the baud rate to the 56 k setting, the process should be over within perhaps 10 min. Monitor progress occasionally during this time. When the update is complete, the hand controller will automatically restart and then prompt you for location and telescope information. Follow the prompts as normal. Your custom data should still be installed. If custom data is absent, you can reinstall your saved data to your telescope by the following process:

From the main menu, select *File/Open*; locate your previously saved data and *open* the file. Drag the file(s) to the *Handbox View* and press *Send*. When complete, power cycle the telescope (turn it off, wait a few seconds, and then turn it on). Check the various menus to see whether any of your data has changed or remains correct. When you have established that everything appears normal, you can then proceed to configure the settings – such as *GPS alignment* – as required. It would be reasonable to recalibrate the sensors and retrain the drives after any firmware update.

Use the Latest Firmware

Installing the latest firmware for your telescope – whether LX200GPS or LX400 – is highly recommended, once you are fully familiarized with the currently loaded version. It should ensure that, at least in most cases, bugs that were identified and reported to Meade (and hopefully fixed!) are finally laid to rest in the update. Never was this more evident than when a number of early updates to the LX400 firmware finally corrected the errant PEC data retention. However, PLEASE remember that you should familiarize yourself with the supplied firmware and *AutoStar* operations for some observing sessions before you tackle the advanced procedure of updating the telescope firmware.

Restoring a Corrupted Handbox

It can happen that during a software upgrade to your handbox, the flash memory used to store programs could get corrupted, leading to a nonfunctioning telescope, or the handbox display may disappear or show garbage characters. This is most likely to occur if the upgrade process is interrupted before completion.

A fresh firmware reload can be initiated as follows: insure that all power to the scope (and handbox) is off. The handbox should be connected to the telescope and the computer with the telescope power off. Power the telescope on and immediately press the 9 button three times (999) on your *AutoStar II* handbox. The message "*Downloading... DO NOT POWER OFF*" should appear. If so, all should be well! Leave the process of reloading alone until it finishes.

Be aware that a new version could bring its own new bugs. You should perhaps wait a few days after a new release in order to find out what the program experts have found when testing the new version.

Free User Software

Andrew Johansen's *PEC editor* and *AutoStar information* pages form an invaluable aid to save (and even edit) the user data in your scope's memory. His application saves *all* user data and does not touch the code. Meade's *ASU* does not save all user data (such as SMT). Andrew Johansen's Web site is http://members.optusnet.com.au/johansea/.

Michael Weasner's Web site is a mine of information about all versions of the *AutoStar* – well worth visiting: http://www.weasner.com/etx/AutoStar_info.html

There is also a commercial updater program *StarPatch* available from *StarGPS*: http://www.stargps.ca/starpatch.htm

This program is more robust than Meade's updater when you have to deal with poor PC-to-telescope connections and even USB drivers, and it also applies patch kits. One user reports that it does not touch stored user data, such as PEC. The software is supposedly free for updating purposes and could be helpful for those experiencing problems with Meade's program.

Telescope Alignment

You will probably have discovered from the beginning, as well as having read the manual, that *Autostar II* asks you to perform a star alignment during the post power on initialization stage. There is a choice of one- or two-star alignment options. When the telescope has been in normal operation with the GPS receiver providing time and location data (resulting in *Autostar II* knowing where it is pointing), the recommended option is the one-star alignment. There are two situations here to be aware of.

Bugs have been identified in the *Autostar II* firmware associated with the simple two-star alignment. Under certain circumstances, this process leaves corrupted data in the *Autostar* software that affects future operations until cleaned up. You can clean up this corruption as suggested by Richard Seymour as follows:

1. Change the telescope setting to *alt/azimuth* aligned.
2. Perform a *one-star* alignment in that mode.
3. Press *enter* when it asks to center the stars.

That process will clean up the corruption. Then set the *Autostar II* back to *Polar*.

Goto Synchronization

During every observing session there will most probably be a nominal error between the position moved to by the telescope after a *goto* command has been issued on the handbox and the precise target coordinates stored within the telescope's database. It is possible to improve future accuracy as a first stage by synchronizing the telescope's encoders to the real target position. After a *goto* has been executed for slewing to a bright star such as Aldebaran, if the target star is off center, center the star in the eyepiece using the *N-S-E-W* keys; press and hold *enter* for about 3 s. The display shows that the handbox is ready to accept the position correction, so press *enter* again; a beep of recognition follows. Later *goto* selections in the same region of the sky should show improved accuracy. Be aware that *goto* errors will accumulate in other regions of the sky. Ideally you should use an illuminated reticle eyepiece for this. There is also a *High Precision Slew* option that takes you to the nearest bright star prior to your selected target, enabling you to center it exactly and therefore improve local target location. This can be useful for identifying some deep space targets. For the next level of accuracy, the scope offers a *Smart Drive* and a *Smart Mount* facility.

Training the Drive

All telescope mounts suffer from some degree of backlash. This is a mechanical defect resulting from an imperfect drive train. Slews in one direction followed by a slew in the opposite direction are not perfectly consistent. For example, if you *goto* an object – such as a well-known bright star on the other side of the sky – and then *goto* the previous bright star, the scope is unlikely to arrive at exactly the same place. Mechanical backlash usually leaves a residual difference along both the RA and declination axes. However, all good mounts provide the means to minimize this discrepancy, and the LX series offers its version – *drive training*. The manual offers a description of the process in the appendix. We recommend using the scope in visual alt-az (altitude-azimuth) mode for this daytime operation, regardless of whether the scope is on an equatorial wedge or not.

Using the handbox, select (from the main menu) *setup, telescope, train drive, az* (azimuth – horizontal) *train*. The handbox instructions then guide you through the process of selecting a terrestrial target and centering it. After completing the azimuth training, repeat the process for the alt(itude) setting. Positions are recorded for drive movements in both azimuth directions and both elevation directions. The differences are noted internally, and the mount subsequently compensates for this. This should minimize the extent of backlash during subsequent slews and should therefore improve pointing accuracy.

The use of a terrestrial target is obviously beneficial! Such targets do not move during the measurements, eliminating any other errors that could confuse the measurements. Meade recommends repeating this drive train process every 3–6 months. It may also be worth repeating the process following any *AutoStar* update in order to avoid the risk of the telescope drive overshooting.

Smart Mount

This is an advanced feature that offers improved pointing accuracy. It requires the user to make a series of specific star position measurements, over a period of between 1 and 2 hours. The process involves activating *Smart Mount* and then centering, in turn, a sequence of some 45 or more handbox-selected stars. Although you can visually center these stars in an illuminated reticle eyepiece, the process can also be carried out using a CCD camera. This process could therefore be carried out later, once a camera is connected via a computer.

Meanwhile, you can still use the visual method to advantage. The manual explains the process. After enabling *Smart Mount*, configure the *AutoStar Suite* planetarium program and set it to follow your mount's position. This way you can see the field anticipated by the telescope and insure that it is matched to the image, thereby avoiding wrongly identified stars. After the scope slews to the first star, you center it, and then press *enter* as instructed. The *AutoStar II* software records any position discrepancy. There is provision for a particular star not being visible due to a physical obstruction: press the *Mode* button, and the handbox selects the next star. Several stars might have to be bypassed during this session if there are environmental obstructions.

There are two points of which you should be aware during this process. The telescope initially selects several bright, well-known stars, but they are usually separated by large distances; this results in your telescope crossing the sky through large movements. Andrew Johansen (mentioned earlier) analyzed the code for this process and explained that the sequence is examining stars within grids that are derived from an algorithm; the actual grid varies with the mount's mode (polar or alt-azimuth), completing two loops – one for azimuth and one for altitude – resulting in the movements seen.

Hopefully all should be well with your SMT project, but you may find that after doing several stars successfully, it drives to the next star but misses it by a large margin. All you can do in this instance is to *mode* to the next star. Should this problem occur repeatedly, there appears to be little that we can do to correct it. study the help file in Andrew's *Beta PEC editor* to learn what could be going on. The editor also displays your SMT model in a grid format, so you can see your end result.

The process usually takes an hour or more to complete, after which the modeled scope can then be saved and subsequently left activated or deactivated. More than one model can be produced, according to your current optical configuration. If you commonly change between two configurations using different balance conditions, you can save the results from these two SMT runs and load the appropriate one when required. *Smart Mount* activation may then improve the telescope's pointing accuracy for you.

Testing SMT

After the process is complete, test the mount using selected stars on the handbox. Hopefully you should at least see something of an improvement in pointing accuracy. Also available within the *Smart Mount* facility is the update procedure. On selection you have to center a selected star; following this, any future sync operation will modify the stored *Smart Mount* database. Again, Andrew's analysis of this process suggests that only stars brighter than magnitude 4.0 will cause the SMT model to be updated. Fainter stars, when synched, will merely adjust the scope's encoder offsets.

Caution, Static!

It is recommended that when not in use, all cables to the front panel should be disconnected. This is to minimize the chance of static electricity damaging the telescope electronics.

Daytime Planetary Viewing

One huge advantage of the LX200 Classic was its ability to be easily synchronized with the sky and the consequent ease of finding the brighter planets during the day. The principle here is that the telescope is always started from a standard parking position. It should be noted that alt-azimuth and equatorially mounted telescopes have different default park positions. The former points due north and at zero altitude in the northern hemisphere, but the wedge-mounted scope points due south and zero declination. Southern hemisphere telescopes point due north when parked.

With the LX200GPS in alt-azimuth mounted mode, in which the base of the scope lies flat on the base of the mount, at power up the scope already *knows* exactly where it is positioned: due north and zero degrees elevation. The motherboard battery (at least in the early models) seemed to insure that it retained the time and the previously entered location coordinates (latitude and longitude). This data is now stored within the flash memory of the *AutoStar II*, and this means that the scope can calculate exactly where it is pointing in the sky. This knowledge results in the handbox *RA-dec* display updating from power on, but only if the GPS setting is *GPS at startup*; otherwise, the time remains unknown. For precision astronomy, the displayed RA and Dec can be confirmed by using the *goto* feature to target a bright star – but never the Sun! One serious point here: as already mentioned, under no circumstances must the telescope be allowed to slew across the disc of the Sun. Such an incident could have the most appalling consequences. Before any daytime *goto* is executed, the front cover of the telescope and the covers on any mounted telescopes – such as the finder – must be securely fitted.

If the planet Venus or Jupiter is visible and sufficiently far from the Sun during the day, the *goto* facility can be used – after taking suitable precautions, as described above and in the manual – and the planet should be seen somewhere in the resulting field of view. This requires only that the telescope should have been powered down via a park command on the previous occasion and is already focused for the eyepiece. You are likely to find it very difficult to focus during the day unless the planet is already near the correct focus position. If a bright moon is visible in the daytime sky, focusing on that is relatively easy. The *goto* feature should only be used when you are familiar with the process and know how the telescope will move. It could be disastrous if the scope drove uncovered in front of the Sun while on the way to locate a planet!

Yahoo Telescope Groups

Yahoo is home to many groups devoted to telescopes, CCD cameras, and other associated topics. The following list shows a number of telescope forums that may be of interest:

http://tech.groups.yahoo.com/group/RCX400/
http://tech.groups.yahoo.com/group/LX200GPS/
http://tech.groups.yahoo.com/group/ASCOM-Talk/
http://tech.groups.yahoo.com/group/Astrometrica/
http://tech.groups.yahoo.com/group/ccd-newastro/
http://tech.groups.yahoo.com/group/guide-user/
http://tech.groups.yahoo.com/group/heq5a/
http://tech.groups.yahoo.com/group/MaximDL/
http://tech.groups.yahoo.com/group/MPOSoftware/
http://tech.groups.yahoo.com/group/starlightxpress/

Also, see Doc G's Information site about LX200GPS telescopes:

http://www.mapug-astronomy.net/ragreiner/index.html

Commercial Servicing Facilities

It is not the intention of this book to try to include features about repairing or even adjusting the actual mechanisms of your telescope. However, when the need for repair or a desire to have a professional astronomer look and adjust your scope arises, the LX200GPS or LX400 user may wish to look into the service offered by Dr. Clay Sherrod of Arkansas Sky Observatory (see Web site URL). Dr. Clay does the scope optimization himself and his service is recognized worldwide. Review the detailed description of the work included in the service schedule (http://www.arksky.org/supercharge.htm).

Summary

This chapter has described the nature of some of the processes, such as balancing, that must be attended to if you wish to get the very best performance from your high-end Meade telescope. Later chapters describe the processes in detail. If you were not planning to do serious imaging with your scope, some of these processes can be omitted, as mentioned in this chapter. Of course, should you change your mind and decide to take your imaging to a higher level, these additional steps should be well worthwhile. In any case, it is worth reading the chapters so that you appreciate the improvements that are possible to get your scope into top condition.

1. Balancing in both axes is essential for successful imaging; the change from heavy eyepiece to camera options possibly involving reducers and different guide scopes will change the whole balance and therefore physical characteristics and operation of the telescope. Using the scope in an unbalanced condition is likely to lead to rapid wear on the gears. See Chap. 4 for full details.
2. Accurate polar alignment is essential for imaging. Alt-azimuth operation, although adequate for visual observing, is not considered a fully effective option for imaging. Depending on where the telescope is pointing, exposures lasting more than even 30 s or so in alt-azimuth mode can lead to star drift and even field rotation. It is possible to use the *AutoStar Suite* to take multiple short exposure images that the software (and third-party software) can stack. Regardless of the software, subsequent image drift depends on the amount of offset between the scope's polar axis and the true celestial pole. A good polar alignment is not difficult to achieve and opens the possibility of obtaining long-exposure images. See Chap. 3 for full details.
3. A good PEC is very beneficial for imaging. Effective guiding is essential, and this is helped considerably by an efficient PEC. The variations in the mechanical precision of your worm train and gears show up as small changes in the position of the target in your field of view. This has little or no effect in an eyepiece but can cause star trails on CCD images. Chapter 6 explains how to greatly reduce, if not virtually eliminate, these errors.

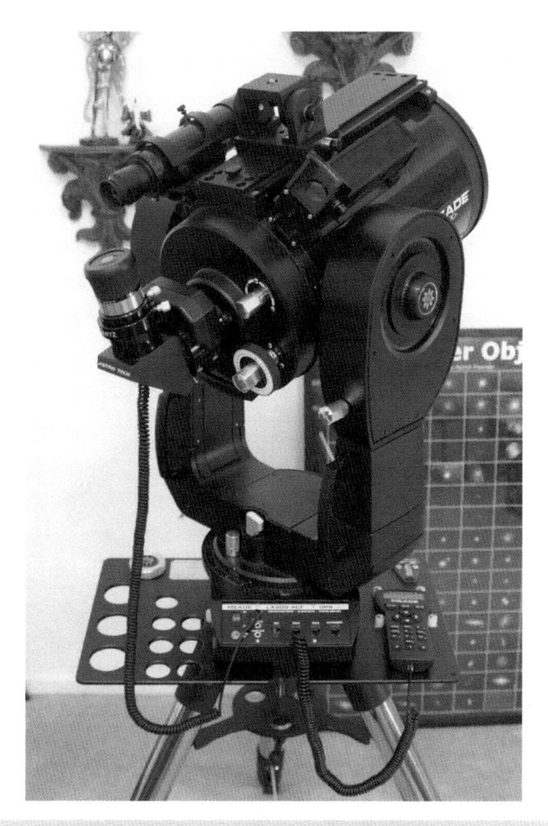

Fig. 2.11 Jim West's customized LX200 showing an eyepiece tray and Telrad finder

4. Reducing backlash by drive training greatly helps autoguiding. The LX series of scopes have a built-in feature to achieve this. The use of Smart Mount should help *goto* accuracy.
5. Synchronizing your scope's actual position (as shown in the eyepiece or on the screen) with that of its current known (displayed) position should improve future pointing accuracy. This helps speed up readiness for imaging.

Jim West has set up his 8-in. Meade LX200-ACF f/10 telescope as shown in Fig. 2.11. He has fitted a scope Tronix Mount Assist Plate, an Astro Tech 2-in. Dielectric Quartz Mirror Diagonal, and uses the Meade Series 5000 Plossl 26-mm eyepiece supplied with the telescope. Further accessories include the ScopeStuff fine focus knob, the popular Telrad finder, and a Losmandy dovetail plate with Losmandy camera mount.

Your telescope is now mounted and some, though not all of the built-in features, have or can be used to the limit of their initial capabilities. You absolutely must balance the telescope in both axes – so this is the next stage on the road to getting the best from the scope.

Chapter 3

Accessories, Great and Small

During the years that the LX200GPS (and associated) series 'scopes have been marketed, a variety of accessories have been designed and manufactured by a number of well-known suppliers, including Meade. It seems that in previous years, suppliers made more profit from selling accessories for Meade 'scopes than they did from selling the actual 'scopes! Figure 3.1 shows a telescope fitted with some of those accessories that are important. Each is discussed in this chapter, along with other useful items, although the detailed applications and benefits are described in specific chapters. Remember, buying the 'scope may just be the start!

Replacement Screws

This unlikely accessory consists of sets of stainless-steel screws for your LX200GPS (and other 'scopes) and seems a worthwhile low-cost addition when you are ordering other items such as a balance rail and weights. With a 'scope permanently mounted in the observatory, many of the small screws in the tube can show signs of rust. It would surely not have been expensive for Meade to have fitted stainless-steel screws from the start. You can swap out several of the screw sets on your 'scope's optical tube assembly.

An Equatorial Wedge?

If you lived at the North or South Pole you could operate your telescope in its basic alt-azimuth setup. You would be able to point your telescope at any star, galaxy, or other comparable target and leave only the right-ascension motor to drive the telescope.

L. Harris, *So You Want a Meade LX Telescope!*, Patrick Moore's Practical Astronomy Series, DOI 10.1007/978-1-4419-1775-1_3, © Springer Science+Business Media, LLC 2010

Fig. 3.1 RCX400/LX400 fitted with balance rail and weights, CCD camera, guide rings and guide 'scope with guide camera, adaptive optics unit with guide camera, and focal reducer in an observatory (Image by the author)

Discounting the effects of atmospheric refraction and drive irregularities, you would be able to follow the object for a long time. Move the telescope to a different latitude and it all changes. You have to activate the declination axis motor to compensate for the drift caused by the latitude change.

The idea of using an equatorial wedge (as it is called) – see Fig. 3.2 and Chap. 4, Fig. 4.1 for examples – is to allow the telescope to be tilted at an angle appropriate to your latitude, so that the telescope can again be driven by just one motor and still keep up with the movement of stars. The support plate of a wedge can be adjusted to compensate for the latitude of your telescope. Because Earth is tilted with respect to the plane of the Solar System, our northern hemisphere sky appears to rotate around the north celestial pole (NCP). In the northern hemisphere, the star Polaris is about 44 arc min away from the true NCP.

Using your telescope without a wedge (that is, using it in alt-azimuth mode, as shown in Fig. 2.11) does not materially affect your viewing of a planet or any other object for many hours. The telescope's right ascension and declination motors will keep the object in the general field of view. The only consequence of this telescope mode is the observed rotation of the field of view. In other words, a planet's disc will appear to slowly rotate clockwise or anticlockwise (depending on your optical configuration and latitude) with the background star field during the session. This

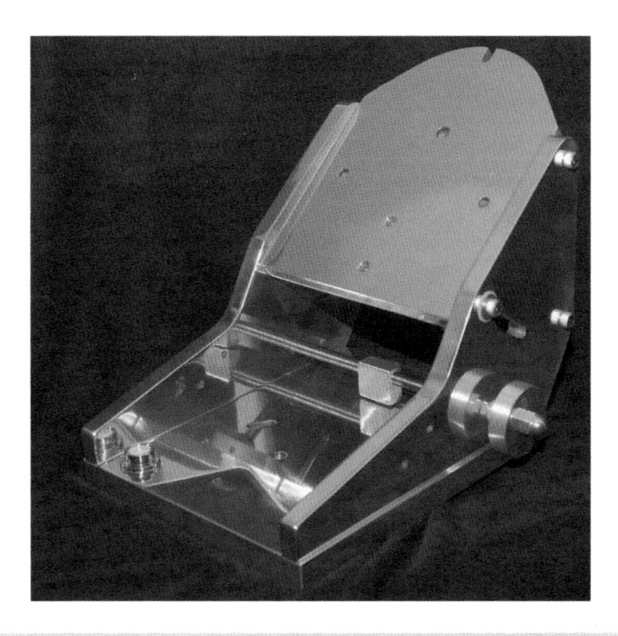

Fig. 3.2 A Milburn wedge, with comprehensive adjustment facilities

does not represent planetary or object rotation on its own axis, but rotation of the field of view itself; it does not usually cause problems.

However, one of the main reasons for buying an LX200GPS/LX400-series 'scope is surely its suitability for CCD imaging. Although adding to the cost of the initial setup, for imaging purposes an equatorial wedge is an essential accessory for those observers not living at the poles! A number of suppliers, including some private individuals, produce these wedges, and all provide full details on the web – see Chap. 2 – so you can peruse the various products. Additionally, there are active forums on the Yahoo website and elsewhere (see references) for most telescopes, including high-end 'scopes. These forums maintain archived messages and data, as well as user experiences in setting up and operating many of the wedges. Wedge prices vary according to the specifications; do be aware that because of the high capabilities of some of the very specialized wedges, manufacture and delivery can sometimes take several weeks.

Meade supplies its own brand of equatorial wedges that are available from a variety of stores and online retailers. You can read a large number of user experiences in the Meade telescope forums and decide how you wish to proceed. Many experienced users have opted for third-party products that offer variations on the Meade product, often including high-quality components individually finished to a high standard. Two of these manufacturers are listed below.

Astro-Engineering UK

http://www.astro-engineering.com
Astro-Engineering is a UK-based company that, as its name implies, manufactures and supplies a wide variety of astronomical accessories. The LX200GPS and LX400 telescopes require fairly substantial mounts with slightly different hole specifications. Early models of the LX400 *MegaWedge Pro* required essential minor modifications to the supporting plate. Models supplied now are ready-to-go. The following is the text of their description:

"*The MegaWedge Pro is our finest wedge for the LX200. Constructed of high precision CNC machined 20 mm thick aluminium plate the wedge is both extraordinarily rigid and lightweight. We designed the MegaWedge Pro specifically to provide not just the last word in rigidity but the ultimate in smooth and precise Polar alignment adjustment. Ideal for use with all (especially the larger) LX200 Classic and new GPS telescopes. The wedge can be field tripod mounted as well as mounted on our Standard, Custom and DIY observatory piers. Equipped with hand slow motion adjusters on both latitude and azimuth registers the MegaWedge Pro is a delight to use and will provide the perfect foundation for advanced astro-imaging projects. Like the standard MegaWedge, the Pro also has a spirit level for fast setup in the field. Finished in black anodic dye and supplied with all fastenings and instructions. The MegaWedge Pro is suitable for latitudes between 63 and 35 degrees.*"

The wedge is straightforward – if somewhat daunting – to use. The realization of the weight of 'scope sitting on that tilt plate can be disconcerting, but adjusting the whole unit is not an ordeal. For smaller, lighter 'scopes, you may not need to buy such a substantial equatorial wedge; the company also makes models for smaller telescopes.

Telescope House

http://www.telescopehouse.com
This UK-based company offers essentially the same range of accessories. Several UK companies distribute similar, sometimes identical products made by the same manufacturers.

Milburn Wedge

http://www.milburnwedge.com/index.php
This wedge, available for the LX200 Classic, is from Ken Milburn *of Bonney Lake Astro Works*. This is one of the top products in the class – see Fig. 3.2. The Deluxe wedge is designed for the heavier models such as the LX400 and the 14-inch (35 cm) LX200GPS. It includes a 25 mm thick base plate, precision spirit level, a

keypad holder, and operates between latitudes 25° and 50° with other ranges available. There is also a choice of finish – either black anodized or silver. The wedge costs approximately $570 plus associated charges, but do check the website for the latest information. The company also manufactures dovetail rail systems, counterbalance weights, camera mounts, and mounting rings.

Balance Rails and Weights

With varying types of hardware, such as cameras, filter units, extra cabling, and top-mounted guide telescopes likely to be fitted to the main optical tube assembly, the telescope's drive motors are likely to be under some strain unless a good balance has been obtained. This is therefore one of the most important accessories that you are likely to buy, and it is not really an optional purchase, unless you only use your 'scope for alt-azimuth viewing. A telescope mounted on an equatorial wedge must be properly balanced in order to insure that the telescope drive motors are not subjected to severe wear and tear. An unbalanced 'scope has very different characteristics from a balanced 'scope; pointing loses accuracy, the motors have to work considerably harder, and the clutch mechanism is under far greater strain. This latter problem can be severe for LX200GPS 'scopes because the declination locking clutch mechanism is not the sturdiest of designs. All these problems are significantly minimized by the careful balancing of the 'scope, as described in Chap. 4.

A rail (or rails) usually comes with a set of small screws (often grub screws) that will match holes already drilled into the telescope tube. Often these drilled and tapped holes have screws already in them, but they are unlikely to be long enough for your rail; they may simply be there for dust protection. The rail package should include fittings that slide along the rail and can be locked, usually using butterfly fittings. Depending on the content of your accessories order (and whether all the components were included), the balance weights should simply screw into the bolts. Complete balance assemblies can be obtained from a number of sources, including some of those listed in the accessories website listing. See Fig. 3.1 showing the rail below the telescope tube.

Dew Heaters

Another near essential accessory for any telescope, especially the LX200GPS and comparable SCT 'scopes, is a dew heater. All telescopes operate in an environment where they are likely to be exposed to cold temperatures for a sustained period. Moisture can condense on any surface where the temperature drops below ambient and reaches dew point, the temperature at which the air becomes saturated with its water content. This is especially prevalent at high humidity levels;

the dew can quickly form across the corrector plate, and within minutes, observing becomes impossible. By wrapping a strap containing heating elements around the perimeter of the corrector plate and passing just enough current to maintain the temperature a degree or two above ambient, dew is largely kept at bay. The Kendrick Dew Remover system includes a controller and circuitry to sense the optimum temperature and allows a controlled current to maintain this small difference. Various controllers are available, so you can select one for your own 'scope's environment.

Dew Shields

A dew shield is another near-essential accessory. Dew shields have the dual roles of providing some light shielding from off-axis light that can otherwise lower image contrast and of delaying the onset of dew by shielding part of the cold sky from the corrector plate. These shields may also be useful for minimizing neighborhood light, but some are not that effective at preventing dew.

The official Meade dew shield is made of metal and is fairly heavy, requiring special attention for telescope balancing when fitted. It is extremely aesthetically pleasing, though! Alternative metal dew shields are also available, and some are made of flexible material. If you use a metal shield, consider buying the Peterson EZ balance weight (see List 3.1), which fits at the back of the tube to balance the shield.

Autoguiding Telescopes

An early accessory in the imager's *must have* list is likely to be an autoguiding telescope – see Fig. 1.5 (Chap. 1) – possibly already in your closet. This is a smaller telescope mounted on top of the main telescope. Fitted with a simple CCD camera (see later), it provides a short exposure image that is analyzed in real time. This provides a reference point that is used to control the exact pointing of the main telescope – using a short timescale feedback. Some caution and careful thought is required for selecting your guide 'scope. Chapter 7 looks at some of the salient points in selecting and setting up the guide 'scope and how you can optimize the parameters. Essentially you have to insure that the guide 'scope's focal ratio and focal length provide a guider image of suitable resolution. The focal ratio insures that the image is likely to have a suitable brightness for best real-time analysis, and the focal length insures that the image resolution is adequate. The author's first guide 'scope was barely adequate for some targets where bright guide stars were too few. The 80 mm guide telescope featured in Fig. 1.5 was purchased later and worked well.

Autoguiding software is normally capable of determining the necessary control signals to command the main telescope to make a small movement to compensate for drift. The guide telescope may add considerable weight to the main telescope, so this must be fully balanced.

To take guided images – as described in Chap. 7 (in which you use a guide telescope to provide a second image that is monitored continuously) – you mount an autoguiding telescope on top of the main telescope (see Fig. 3.1). The LX200GPS (and other Meade 'scopes) has sets of holes predrilled on both the top of the 'scope (for this purpose) and below for mounting a balance rail. Mounting a guide 'scope requires two mounting rings that are fitted to a mounting rail. Figure 4.7 in Chap. 4 shows the fittings where the long guide 'scope adjusting screws allow a wide range of guide 'scope diameters to be used.

Focal Reducers

Meade telescopes are normally manufactured to operate at f10 (except for the LX400, which operates at f8); a 300 mm diameter Meade mirror therefore has a nominal focal length of 3,000 mm, and pro rata for other mirror sizes. The LX400 (f8) 300 mm diameter mirror has a quoted nominal focal length of 2,438 mm. This works well for many applications. It is also excellent for visual use with the Meade-supplied eyepiece. However, for most imaging applications, especially nova hunting and astrometry, you need to use the widest possible field of view (more stars equals higher chance of discovery). It is a common practice therefore to fit a focal reducer, sometimes called a *field flattener*. Meade produces its own reducers, currently an f3.3 and an f6.3 reducer. These are designed for the LX200GPS series telescopes, though a Celestron f6.3 reducer can be used on an LX400 'scope. Some other manufacturers have produced reducers designed for SCT 'scopes (Schmidt Cassegrain telescopes, such as the LX series), and their performance is often discussed on the LX forums.

Focal reducers must be positioned on the back of the 'scope as the first accessory. This is a significant restriction on positioning when used with SCT telescopes but is essential because of the precision design of the reducer optics. Such reducers have components specifically designed to match the SCT design. Focal reducers are added to the optical path and must be used according to the instructions provided with them. It is essential that they are used close to their design distance from the imaging plane, and it is not possible to stack two reducers together. The basic telescope produces a real image approximately 10 cm beyond the rear of the tube opening, and this can normally be focused on the CCD camera plane. The positioning of the focal reducer brings that real image closer to the back of the tube, so there is a limit on the amount of further equipment that can be interposed without preventing a focus from being achieved. Focal reducers are popular because they make for brighter images by reducing the image size. This is achieved by reducing the focal ratio, for example from f10 to f6.3.

The reducer's best optical performance is obtained when used at the design distance, although there is some tolerance for use at other distances that will produce slightly different reduction ratios. When focused at infinity, the normal position of focus when used at f10, the real image is already close to the back plate of the telescope. If other items such as reducers or tubes are added, the telescope's mirror must be moved forward in the tube to push the real image backwards. There is a maximum distance that can be accommodated this way, known as the maximum back focus distance. Consequently, there is a limit to the amount of focal reduction that can be achieved using reducers. The LX400 series 'scope uses a fixed mirror; the focus is adjusted by moving the secondary mirror, and therefore the correcting plate, backwards and forwards within the tube. Clearly, again, there is a limit to the amount of additional equipment that can be added to the rear of the tube if you are still to be able to focus the 'scope properly. The use of filter wheels, and even flip mirrors, may therefore not be possible with some reducers. Wide experience has indicated that the standard f6.3 reducer still offers the opportunity to add other equipment, perhaps including an adaptive optics unit incorporating a filter.

An inherent problem with most reducers is the production of an image with vignetting. This is not necessarily caused by the reducer itself; it can be a consequence of the inherent design of the telescope. Reducers have a front diameter and an exit pupil. Only light rays that can leave this pupil will reach the image plane. If the initial image through the telescope is not uniformly illuminated over its full surface by the cone of light from the telescope, then neither will the reduced image be fully illuminated. This is not a significant problem with a properly designed focal reducer used at its design distance. However, if a large field CCD chip (particularly one the size of 35 mm film) is used, the telescope will not normally illuminate the full image and therefore will not fully illuminate the reduced image. Using the 35 mm example, with an f6.3 reducer the original circle of illumination would need to be about 70 mm if none of the image is to be lost – far larger than the normal exit opening of most small SCTs. Consequently, some degree of under illumination of the outer concentric regions of the image plane is likely to be experienced. The acceptance of a smaller effective reduction should produce less vignetting. By always taking flats prior to or following any observing session, you can help compensate for uneven illumination to a large extent. Flats are calibration images, simply a uniformly illuminated field viewed by the telescope (see Chap. 5, Fig. 5.8).

Barlow (Extender) Units

Focal magnifiers do the opposite of focal reducers – they extend the effective focal length of the 'scope in order to produce a greater magnification. They are often used for planetary viewing and imaging where the greater magnification is advantageous.

Just as a reducer produces a brighter image than the telescope when used at its native focal length, so a magnifier produces a larger image having correspondingly less inherent brightness. The amount of magnification produced by a focal magnifier depends on the spacing between the magnifier unit and the focal plane. Magnifiers and reducers are designed to work at a specified magnification/reduction ratio and therefore at a specified distance from the image focal plane. However, their positioning is not quite as critical as that of reducers. Commonly used Barlows are operated at magnifications between two and three times the 'scope's normal focal length, making them popular with planetary imagers because such imaging requires the highest feasible resolution to show the tiny discs of planets.

CCD Cameras: Main and Guide

The range of CCD cameras available to the amateur imaging enthusiast has increased considerably in recent years. You can buy general purpose imagers at fairly low cost for a basic model, but many enthusiasts have opted for a more advanced unit from one of the main suppliers. The Santa Barbara Instrument Group, Starlight Xpress, Atik Instruments (Artemis cameras), QSI, and Meade are all producing cameras for the astronomer. Camera suppliers invariably include software to control the camera. Michael Barber of SBIG says, "Virtually all of our cameras come with *CCDOPS*, *CCDSoftV5*, *TheSky version 5,* and *Equinox* (for Mac computers on request at no extra charge)."

Prices range enormously because of the huge differences in capabilities and sensitivities. They can be arbitrarily divided into two camps – those that use the very low noise Sony chips and those using Kodak and comparable chips, often having shutters built in. For anyone contemplating buying a high quality CCD camera, look carefully at the range of cameras, including one-shot color cameras that have a built in Bayer matrix over the chip.

For the serious imaging projects discussed in this book, you are likely to want a main camera and a guide camera. In some models, these are combined, but usually they are two separate cameras, with the guide camera being of considerably smaller size and lower performance. The main requirement for the guide camera is to be able to produce a guide star in most imaging situations. One popular choice is the *Starlight Xpress H9C* one-shot color camera – see Fig. 3.3 and the small autoguide camera. A more recent guide camera, the Lodestar, is shown in Fig. 3.4. Both cameras offer exceptionally low noise levels. In virtually every camera, the active chip is cooled, sometimes in two stages. The cooling drastically reduces the thermal noise inherent in all such devices. On average, the chip is maintained at about 30°C below ambient temperature.

Recently, a number of large format chip cameras have been produced, even exceeding the old 35 mm format. The larger sizes are probably susceptible to severe vignetting if used with telescopes using inordinately powerful focal reducers. An f3.3 reducer fitted to an LX200GPS combined with a large format CCD camera could lead

Fig. 3.3 Starlight Xpress SXV-H9C one-shot color camera. This can be used for deep sky imaging as well as astrometry

Fig. 3.4 Starlight Xpress Lodestar guide camera. This is a high-quality guide camera that could even be used for normal imaging

to potentially disappointing results, so be aware of the need to check before purchase. It could, of course, be possible to partially compensate for such vignetting by using suitable flat frames. These are described in a later chapter but are essentially calibration images taken of a uniformly illuminated background as the reference image.

Filters

As astronomers, we do have a secret or two! There is a lot of discussion about light pollution and the need to minimize the wastage of power, and to enable everyone to enjoy the night sky. The imaging astronomer can fight back to some extent by using light-pollution filters. One option is the IDAS LPS-P2 series of light-pollution filters. These are widely available and not too expensive. They are designed to suppress the common emission lines generated by artificial lighting, while allowing important nebula emission lines to pass, thus enhancing the contrast of astronomical objects, particularly emission nebulae. If you take some unfiltered one-shot color images through your LX200GPS 'scope (or similar), you will almost always find a color cast caused by light pollution. These filters provide a balanced color transmission. Do be aware that all filters attenuate the light reaching your eyes or CCD chip. That is, they do not make objects brighter. Rather, they increase the contrast over the background by significantly reducing polluting background light. Filters, particularly the IDAS LPS-P2 filter, come with their own transmission characteristics. Such filters are carefully designed to attenuate only specific emission lines associated with light pollution.

Software

Although some of the big improvements to the performance of your LX200GPS telescope can be accomplished using the software and features available for your telescope, you are really going to want to use third-party software to raise the efficiency of several parameters. This is where you depart from the purely visual experience of using the telescope and enter the world of CCD imaging, for which the LX200GPS and LX400 telescopes were eminently designed.

Long exposure imaging cannot be done without suitably sophisticated software, and there is a reasonable choice available. All CCD cameras come with software normally capable of downloading images from the camera and providing basic image processing options. However, additional software can greatly enhance the whole imaging experience. *Maxim DL* is good for telescope and camera control, but *AstroArt* can be used for camera control and processing alone, while using the Meade handbox for telescope control. This book includes several chapters covering essential and advanced software, featuring *ACP* and *PEMPro* (both to be described later), together with the *ASCOM* platform that is freely available. You might also consider using *Guide-8* as your planetarium program. Some other programs, such as *Sky*6, can be used for both telescope and camera control. Before purchase, do check that any software is fully compatible with your own hardware. Some software that may appear comprehensive may not actually control your Meade telescope, or perhaps not your chosen CCD camera. See the list of websites for links to suppliers' sites and product costs.

Adaptive Optics Units

Chapter 9 is devoted to the use of adaptive optics (sometimes called active optics) devices. These units – see Fig. 3.5 – were developed from the application of devices originally devised by professional astronomers. Within the device, a sample bright star image very close to the imaged field is collected and analyzed, and all changes of position caused by high-frequency atmospheric scintillation are compensated by the instantaneous control of a corrective tilted plate that adjusts the position of the whole image to compensate. The resulting image, when used in a fully calibrated unit, should have the best possible image quality. As would be expected, these devices are not cheap. Chapter 9 should help you decide whether one of these units would be of interest to you. There are a number of manufacturers of these units. Santa Barbara Instrument Group and Starlight Xpress both produce very popular cameras.

Observatories

Your LX200GPS/LX400 is a sophisticated telescope, and you naturally want to set it up to work at maximum efficiency. You are also likely to want to do as much observing as possible without having to spend half an hour or so aligning the

Fig. 3.5 A Starlight Xpress adaptive optics unit fitted on the LX400 telescope. Note the need for extra control cables

telescope before every session. Many amateurs therefore opt to build or buy an observatory. The choice is enormous; a number of astronomers use basic sheds very successfully, making modifications such as easy roof removal, possibly incorporating a sliding rail system. Decisions about the size, type, and location of such an observatory can take some time, and cost considerations can limit you to using a cheap shed for the purpose. If you have carpentry skills, you can probably make some modifications to such a shed and develop a workable design, saving you large sums of money. Above all, talk to fellow astronomers – especially those with observatories. They are always ready to chat about the problems and joys of having your equipment in your own observatory. There are several websites showing amateur observatories of all types; run-off shed roof observatories are probably the most popular (see Fig. 3.6). While the telescope is covered, it has full protection from the elements. The run-off roof can be easily removed by sliding it along the rails, enabling the 'scope to have wide access to the sky.

Several years ago, the author finally had the garden space to build an observatory. What follows is a personal account of event. "The choice of sizes and makes appeared to be rather limited, but I was happy to finally order a fiber-glass observatory. Having done so, the location of the observatory and ground preparation was paramount. By carefully checking the position where tree and neighborhood roof interference was minimal, I staked the plot. Before ordering, I checked the preliminary requirements for the concrete base and also that the observatory sections could be taken through the side entrance to get them into the garden. I then

Fig. 3.6 Run-off shed observatory (Image courtesy of Hilary Jones; http://www.muffycat.org/astronomy)

arranged delivery to the house. I wanted an electricity supply to be available within the observatory, so I prepared some conduit piping and dug a deep trench to allow a main supply armored cable to be run under the base and up into the structure. I had a builder provide foundations and mix and lay the concrete. It then became evident that the supplier had given me the wrong base dimensions in the instructions for assembly. The result was that my wife and I had to visit the local hardware store to purchase and carry several bags of ready-mix concrete and commence extending the base! This was eventually completed, although the result was not entirely satisfactory. I can recommend that you insure that you know the precise requirements and then allow some more, particularly making sure that you know the *external* diameter of your observatory! It is also advisable to seal the surface of your base, using a suitable concrete sealant to minimize dust production.

The main components of the observatory were laid across the grass near the concrete base. There are four wall sections each covering a 90 degree curvature, including the door section and two large dome sections plus a shutter section to which the two hemispheres were to be bolted. The walls of the structure are bolted together forming the familiar circular appearance. This structure should ideally sit on a cushion of flexible waterproof material that keeps rain outside, rather than allowing it to seep under the walls. Such a cushion need only be a few millimeters thick. The dome with shutters is also assembled and the supplied wheels fitted to the dome base – see Fig. 3.7 – and then greased. The entire weight of the dome required a group of friends to be called upon from the local astronomical society. The dome was ceremoniously raised onto the upper edge of the walls and carefully lowered into position.

Fig. 3.7 Completed dome ready to mount on circular walls

After checking that it rotated freely, friends were thanked and the telescope components then carried into the new observatory. The concrete base was covered with carpet tiles, to further reduce dust.

Once installed, whatever form your observatory takes, you are likely to be amazed at the benefits experienced. The 'scope can be carefully set up and polar aligned once and for all. Equipment can often be left fully installed and connected, though be aware of possible weather effects, such as severe condensation on occasions" (see later paragraph).

Following the installation of my telescope inside the new observatory, the first observation – *first light* – was that of the planet Jupiter. Instead of the shimmering image always previously experienced outside while exposed to the elements, the image was rock steady. This was a significant difference; gone were the minute oscillations caused by breezes to which one can become used over the years. Even moderate winds had little effect on image stability. It was possible to set up my computer with ease; a computer desk meant that a session could be up and running within minutes. Before observing, I now go out and power up the CCD camera about 30 min in advance to allow stable cooling to be achieved. My concession to the neighborhood is that after 10 p.m. local time I slowly and carefully rotate the dome as silently as possible. No one can ever complain that astronomy is noisy! Books have been written about building and using amateur observatories, so you might wish to consider researching this further before committing yourself financially".

A more recent observatory designed by Farhat Hanna and Wayne Parker is the *SkyShed Pod* (see website list). This innovative design – see Figs. 3.8 and 3.9 – comprises six wall sections and four dome sections that together are apparently portable enough even to take to star parties. Many variations in style and color are

Fig. 3.8 SkyShed POD (image courtesy Ian King Imaging)

Fig. 3.9 SkyShed POD (image courtesy Nick Witte-Vermeulen of Altair Astro, UK)

available. First time assembly is described as taking less than 2 h, including the assembly of rain seals and wheel tracks that are bolted into place. The product is 2.4 m diameter at the widest point, and as with other observatories, bolting to the ground is recommended. Due to the dome's design, another advantage of the Skyshed is that the dome does not need to be constantly rotated when your telescope tracks the sky. Movement is only needed every few hours. The designers claim a faster telescope cool down time due to the wide opening. Although clearly ideal for situations where more than one person is present in the observatory, An aperture option is currently being developed to offer better protection from the wind.

Dehumidifiers

An additional and unexpected cost was that of a dehumidifier. To reduce the incidences of severe condensation coating everything, you should consider buying a family-size dehumidifier that you can leave running as appropriate. You can switch this on at the end of an observing session when you leave the observatory. You then know that all will be well the following morning. The unit can be controlled by a time switch when it is not considered necessary to leave it on continuously. The dehumidifier does not operate below about 5°C, so you might sometimes use a small fan heater set to its frost setting. This allows gentle warmth to circulate within the confines of the observatory and automatically switches off above about 6°C, leaving the dehumidifier to dry the air while also gently warming it.

Hartmann Mask

The design and use of the Hartmann mask was included in Chap. 2. This particular accessory is almost essential yet not too difficult to produce as a home-made item. Use the type of filter in which one of the holes is threaded to take a solar filter for white-light viewing, which allows this to serve as a dual-use mask.

Bahtinov Mask

The Bahtinov mask is a more recent innovation, devised by amateur astrophotographer Pavel Bahtinov It is used as a focusing aid and consists of a disk made from opaque material fitted over the end of the telescope. Unlike the circular holes of a Hartman mask, this has slots cut out in a specific pattern – see Fig. 3.10. The website (http://astrojargon.net/MaskGenerator.aspx) provides a design for your specific telescope after you input your telescope's basic parameters. There is also an additional feature to optimize your resulting design if you so wish.

Fig. 3.10 Bahtinov mask courtesy Nicolaj Haarup

List 3.1 Websites for LX200GPS and associated accessories

www.astro-physics.com/
www.scopestuff.com/
www.astro-engineering.com/
www.iankingimaging.com/
www.telescopehouse.com/
www.petersonengineering.com/sky/index.htm
www.skyshedpod.com/
www.kendrickastro.com/astro/dewremover.html
www.widescreen-centre.co.uk/
www.bisque.com/help/v6/TheSky_Version_6.htm
www.sbig.com/
www.cyanogen.com/
www.msb-astroart.com/
www.starlight-xpress.co.uk/

Chapter 4

Balancing and Polar Alignment

Wedges, good balancing, and good polar alignment are fundamental to the serious imager. In this chapter, essential wedge features for easy adjustment are discussed, and hints are given about how to set up a wedge prior to mounting the telescope. The importance of accurate balancing and how that can be achieved are also discussed.

Polar alignment using only the original telescope and supplied optics is covered, and a suggested target for good polar alignment is given. The methods of adjusting the wedge azimuth and latitude are explained in detail, and hints are offered on how to make the process easier. Recent methods of measuring alignment errors in software are discussed.

Equatorial Wedges

Although, as discussed in the previous chapter, a wedge is not necessary for purely visual use, it is almost essential for long-exposure imaging. Converting your basic telescope into a polar-mounted instrument forms an essential upgrade for anyone proposing to take long-exposure images. If you do *not* plan to do long-exposure imaging, a more cost-effective telescope having some comparable features can be bought, as discussed earlier. Your LX200GPS/LX400 is an extremely capable 'scope; let us see what you need to fulfill its potential. (See Chap. 3 for full details about considering equatorial wedges, as well as Fig. 3.2 and this chapter for examples, and List 3.1.)

Figures 4.1 and 4.2 show different views of an equatorial wedge. Other top-end models should have comparable features. The bubble level can be seen on the left-hand side of Fig. 4.1 on the platform section. Note that this level provides an approximate guide only; it is not meant to be a precision feature. By adjusting the sturdy tripod with the wedge fitted, a reasonable tripod level can be achieved. To improve accuracy, you can place a builder's level on the base and check and adjust it with that.

L. Harris, *So You Want a Meade LX Telescope!*, Patrick Moore's Practical Astronomy Series,
DOI 10.1007/978-1-4419-1775-1_4, © Springer Science+Business Media, LLC 2010

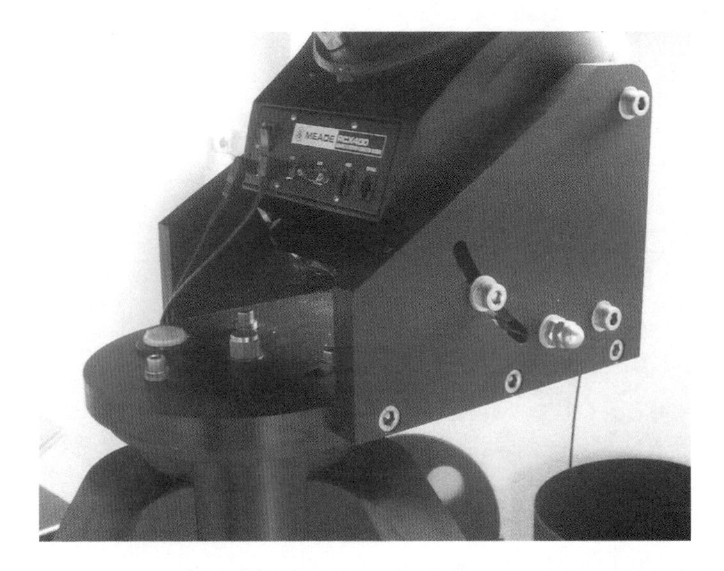

Fig. 4.1 Telescope mounted on an equatorial wedge, showing side bolts and latitude adjustment slot

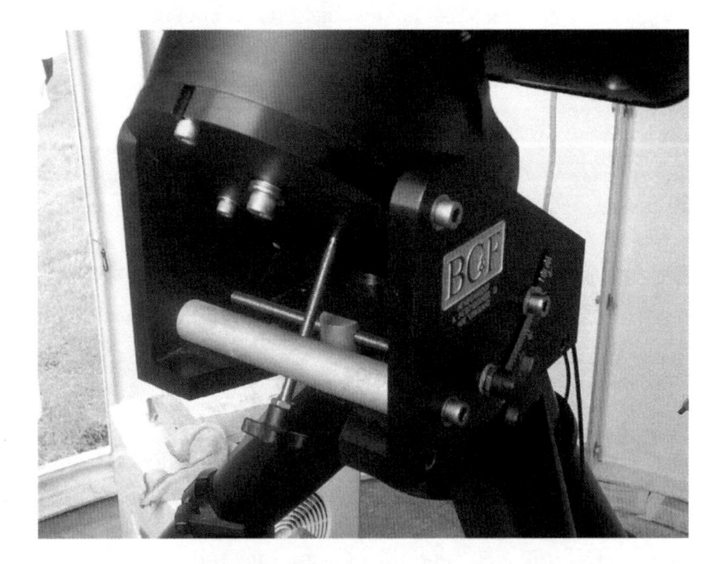

Fig. 4.2 Rear of wedge (the view from the south side) showing a latitude adjustment bolt

The wedge base is bolted to the tripod using three standard bolts arranged at 120° to each other, and there is also a large central bolt fitting through the tripod (see Fig. 4.3). The side of the wedge seen in Figs. 4.1 and 4.2 shows three bolts on the bottom edge fixing the base plate on this side; there are corresponding bolts on the right-hand side as well. A black adjustable knob can be seen on the side of Fig. 4.2; this knob is used to move the azimuth plate relative to the tripod base.

Fig. 4.3 Upper face of wedge showing bubble level and bolts securing wedge to mount

The three outside bolt heads and one central bolt head that hold the wedge body fixed to the tripod head *must* be slackened off before turning the azimuth adjustment knob, in order to permit the necessary movement. The central latitude adjustment bolt can also be seen in Fig. 4.2 and supports the weight of the tilt plate that carries the telescope itself. This long bolt is threaded through a cylinder diameter that is held rigidly by two bolts on either side of the wedge. These two bolts should be slackened off before making latitude adjustments; this can then allow the cylinder to rotate slightly during adjustment. The telescope itself is bolted to the tilt plate using the three bolts provided with the telescope, as seen in Fig. 4.2. The latitude setting is obtained by slowly turning this knob until the tilt plate indicates your latitude, but read the following text first before making any adjustments.

Setting Up Your 'Scope

The basic fitting of the initial components to your telescope will be fully described in the telescope's manual, and so we have not included it here. However, if the telescope is large and heavy, you need to do some planning in advance of its construction. You should identify where you are going to install it. The author's first and second Meade LX200 (Classic) telescopes were 10 inches (25 cm) and 12 inches

(30 cm) diameter, respectively, barely capable of being lifted by one or even two people. Before constructing the telescope, you have to identify the best position for the observatory, and this requires carefully noting the angular height of neighbors' trees. When an observatory location is finally identified and the observatory built, the telescope components are taken inside it and assembled.

Unless you plan to routinely dismantle the OTA from the tripod mount at the end of each session, an observatory offers the huge advantage of permanency as well as much improved ease of operation and likely increased frequency of use. Let us assume that you have decided exactly where your telescope is to be positioned. The heavy-duty tripod has been positioned and the wedge has been fitted. Before continuing, the wedge must be set up correctly for operations.

Getting an Approximate Polar Alignment

There are several ways to achieve a rough polar alignment. Wait for a sunny day, but know in advance when the Sun will be exactly due south. You can find this out online, or better still, from a good planetarium program such as *Guide-8*. When set up with your correct time zone, latitude, and longitude, *Guide* (and many other comparable programs) can tell you exactly when the Sun will be due south – at 1153 UTC on this particular day. The exact time of solar transit varies throughout the year. In advance of this time, you should set up suitable markers to allow a shadow to define the north to south alignment. This direction is then identified in the observatory, enabling you to position the tripod and equatorial wedge to within a degree or two of the meridian. A clear evening sky enables you to confirm that the alignment is about right (by viewing Polaris by eye).

Although Polaris (the North Star) is not exactly at the north celestial pole (NCP), there are two occasions during the night when it is either directly above the meridian or directly below it. Again, *Guide-8* (and other programs) can show this graphically. This simple method should allow you to align the azimuth of your telescope mount within a degree or two of the true north celestial pole, which will help enormously when you adjust the azimuth controls.

Next the latitude: by checking your latitude using either a local map, or better still a GPS receiver, you can adjust the latitude setting of the equatorial wedge. An important hint here: err on the side of positioning your polar axis slightly higher than your best estimate of your latitude. This will always be an approximation because these settings can only be read to about half a degree, perhaps one quarter of a degree at best. If you leave the axis slightly higher – possibly as much as one degree – than your best judgment, you will find elevation refinement, discussed later, much easier to perform! It is considerably easier to *lower* the polar axis with all that weight on it, than it is to *raise* it!

At this stage, with the equatorial wedge approximately aligned but securely bolted down, you can arrange for the heavy OTA to be mounted. The 'scope is then ready for the next stage – balancing – then to be followed by accurate polar alignment.

Chapter 2 explained the recommended power on procedures and the first jobs to be done during these initial sessions, both daylight testing and the first night-time session.

Balancing in Declination

The importance of having a well-balanced telescope is not always obvious to the novice user. By well balanced we mean what happens when the declination clutch and the RA clutch are released. When the declination clutch is released, a well-balanced tube should have no tendency to move either up or down, no matter what declination the telescope tube is currently positioned in. It is sometimes considered good practice to have one end, usually the eyepiece end, *slightly* heavier than the front. This ensures that the declination axis drive worm remains fully engaged with the teeth. Otherwise, if a truly perfect balance was obtained, there could be occasions when the teeth remained out of contact (floating), possibly causing oscillations. Similarly, the RA axis can be balanced so that there is a *slight* tendency for the 'scope to move eastwards, ensuring that the teeth remain constantly held by the worm drive (which is normally driving the 'scope westwards).

If the 'scope is not properly balanced, a number of not so obvious factors can affect performance. It becomes necessary to tighten clutches with far more torque than would otherwise be necessary to hold the tube. This can rapidly lead to mechanical damage and is very likely to prevent the telescope from driving accurately. Wear and tear on the gears and other components could be excessive.

Declination Axis Balance

To obtain declination axis balance, a rail carrying weights is usually fitted to the underside of the 'scope (see Fig. 4.4). The picture shows the rail at the end of the balance procedure. It is possible to fit the rail on the top of the tube, but this is usually where you would mount a guide telescope, and that would limit the balancing capabilities. If you are mechanically minded you might wish to cut, drill, and fit your own balance rail. Without such tools, you can buy these fittings from established astronomical equipment stores (see Chap. 3). As a subscriber to the LX200GPS and LX400 forums on Yahoo, you can become familiar with a number of specialists who have developed their own series of products, many of which are especially for the LX200GPS and LX400 series telescopes. As an owner of one of these 'scopes, it could be well worth the time monitoring the forum posts to appreciate how useful several of these products are. For example, you might want to order an LX400 balance rail and a selection of weights. Choosing sufficient weights is not easy because you cannot always anticipate how your long-term balance requirements will develop.

There is a logical approach to balancing your 'scope. The balance procedure largely applies to the equatorially mounted 'scope, not those used in alt-azimuth mode.

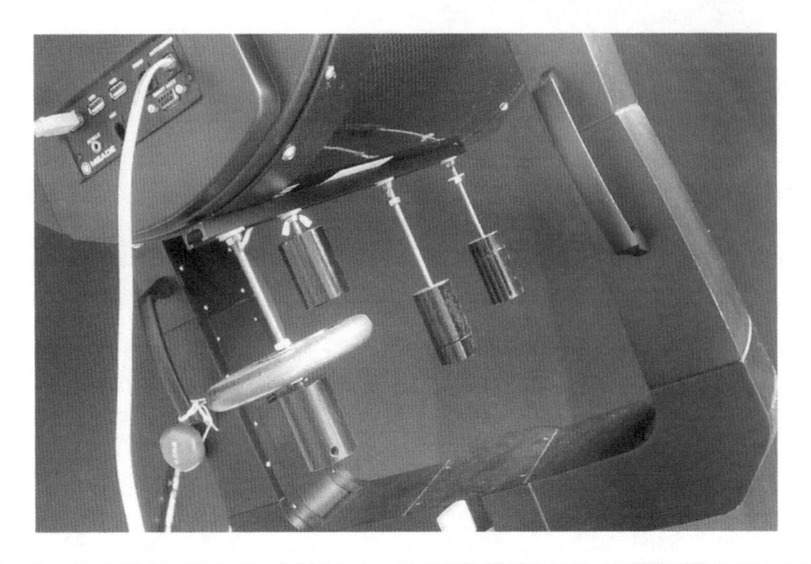

Fig. 4.4 Rear of LX400 showing rail fitting with various weights on different rod lengths

Although you might wish to jump straight in and fit your 'scope with your most commonly used configuration, it could be better to pause and consider your likely long-term needs.

This author's normal telescope configuration includes the following:

1. an 800-mm guide telescope mounted on top;
2. an F6.3 reducing converter;
3. a combination *Starlight Xpress* adaptive optics and SXV-H9C CCD camera unit.

In case the above combination seems curious – a separate guide telescope despite having an adaptive optics unit that has its own guide unit built in – the OTA-mounted guide telescope is used for those targets where no suitable field-of-view guide star has been identified.

If you are ready to obtain your first balance, you should find it much easier without a guide telescope adding to the tube weight. Figure 4.5 shows the extra weight required to balance the 'scope in declination when the guide 'scope is fitted. You are likely to want a guide 'scope sooner or later for imaging, but if you are balancing the 'scope for the first time and are not yet fitting a guide 'scope, this can be left off during initial balancing. A three-dimensional balance will be required to include both RA and declination axes.

Vertical Balance

The topic of whether you start by doing a vertical or horizontal balance has been discussed on the forums. Some favor one method, some favor the other.

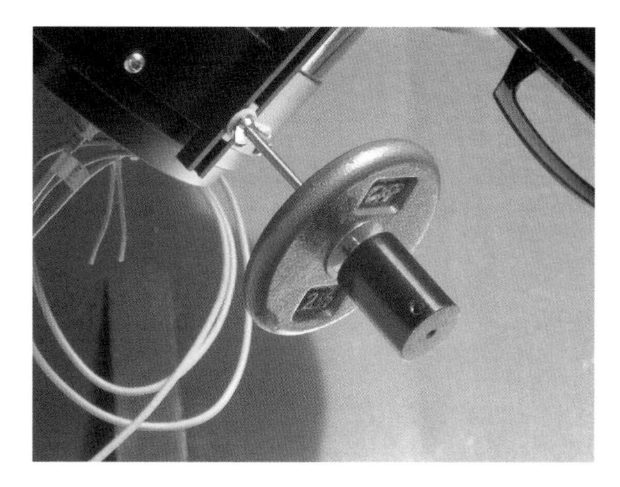

Fig. 4.5 The additional balance weight required when adding a guide 'scope

This author's preference is always to do the vertical balance first because this orientation defines the total weight required for balance. The heavier your guide 'scope, the heavier the balance weights that will be required. Either way, you are aiming for a complete balance at the end of the process.

The first balance should be in the vertical position, in which accessories placed on the top of the 'scope are counterbalanced by weights added to the bottom rail (see Fig. 4.6). Clearly the absence of a guide 'scope would require significantly less balancing weight than shown here (see Fig. 4.7). Start by fitting some weights close to the OTA. You can position the telescope vertically with the declination clutch released, but hold the tube. If you have a guide 'scope fitted, the tube will almost certainly be completely unbalanced, causing a strong tendency for the vertical tube to swing in the direction of heaviest weight. Without the guide 'scope, balance would be much closer. With the OTA still held in the vertical position, you can see which way the counterweights need to be adjusted. Add weights to the lower side of the OTA. One or more of the weights can be fitted on the end of longer rods, shown in these pictures, to increase the range of balance capability. Each weight has a greater effect (its moment) when held further from the tube on these longer rods, especially during the vertical balance process.

Insure that the extended rod has clearance for the base of the fork arms or, if not, be aware of this situation. This weight should not be the limiting factor for the camera being able to reach high northern declinations. The long supporting threaded rods seen in Figs. 4.4–4.6 enable the weight(s) to be moved away from or toward the tube. As a balance point is neared, you will sense this and realize that you can get there eventually.

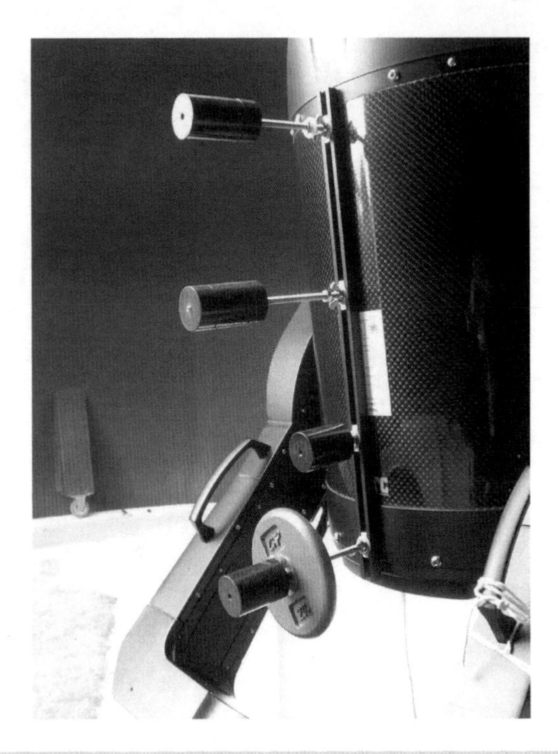

Fig. 4.6 Balancing the LX400 OTA in vertical position

If you have heavy equipment, for instance a metal dew shield on the front end of the 'scope, this cannot easily be balanced because of the movement of the center of gravity. A Peterson rear cell balance could be the ideal answer if you have an LX200GPS-series 'scope (see Chap. 3). Alternatively, you may have to add weights to the rear handles on the OTA. This may seem inelegant, but anything that results in a really good telescope balance is highly desirable. Instead of fitting a metal dew tube, try using a sturdy felt-lined tube bought from Kendrick, which weighs much less than the metal equivalent.

By adjusting the weights and their position, and then rechecking, you should be able to obtain a good vertical balance. The weights seen in the pictures are shown at their final (balance) positions. This balance will only be valid for a swing of a few degrees from vertical because you have not yet done a *complete* balance. At the point of vertical balance, the weight on the side of the tube where any equipment is fitted is balanced by the weights below the tube. At a later time, when you have added a guide 'scope on the top of the OTA, significant additional weight will be required on the underside to rebalance the 'scope, as seen in Fig. 4.5.

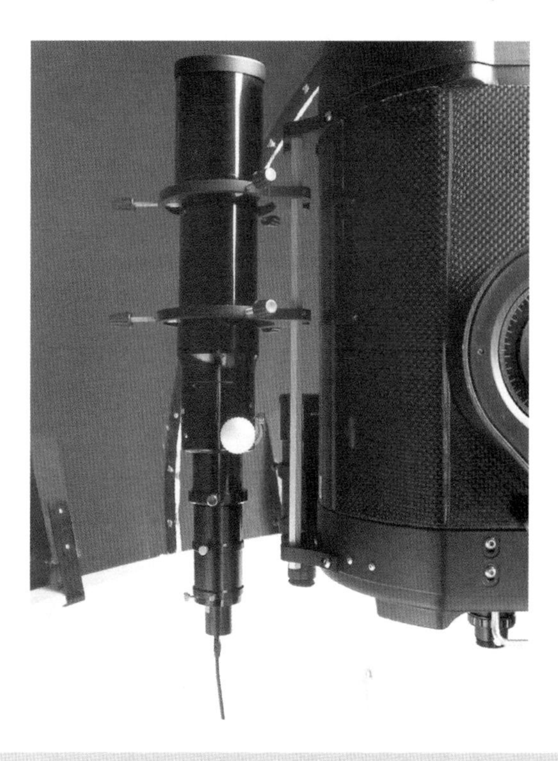

Fig. 4.7 A heavy guide 'scope requires extra balance weights

Horizontal Balance

The next job is to balance the telescope horizontally. Again, hold the tube horizontally with the clutch released. The 'scope will try to either rise or fall dramatically; this tendency indicates the amount of strain that would be experienced by the clutch without proper balance. You are aiming to reduce the strain to almost zero. If the front end of the telescope drops, the weights need to be moved toward the rear of the 'scope – and vice versa if the 'scope tilts upwards. Move the weights along the rail to improve the balance. Depending on the type of 'scope and the weight of accessories (combinations of guide 'scope and CCD camera), a significant weight may have been required for perfect balance. If you have an LX200GPS, the Peterson rear cell balance referred to previously can be bought and fitted at the mirror end, allowing a full-weight dew tube to be added. It is useful to have a number of separate weights on the rail, rather than one really heavy one, in order to help spread the load along the rail. When the tube balances properly in the horizontal, you should then repeat the vertical balance checks. Check and then recheck the horizontal balance; you should be able to achieve a perfect declination balance, and this will in turn greatly aid the control and pointing accuracy of the 'scope. Next, we will balance the RA axis.

Right Ascension Axis Balance

This process is rather simpler than balancing in declination, especially if you already have a suitable weight. But first, note that there has been some discussion about this particular aspect of balancing a 'scope. Some experts have suggested that even though the native 'scope is westward heavy (because of the declination drive motor on the west arm), that this lack of balance can be accepted regardless. Others suggest, including this author, that it is better to at least partly compensate for the imbalance.

We are therefore going to improve the balance of the fork arms by placing additional weights on the left fork arm (the one without the declination motor). There are several ways to do this. Release the RA clutch while steadying the forks; it will be very surprising if the 'scope does not quickly lurch to the west. With extra hardware built in to the west fork arm, your 'scope is always likely to respond this way, so you must fix balancing weights on and possibly within the east fork arm.

The aim is not only to limit the westward movement tendency but to possibly encourage the 'scope to have a tendency to drift to the east. No suitable balance weight system that can compensate for the 'scope's natural westward tendency is currently easily available commercially. One recommendation has been to put lead shot (or something similar) into small canisters and place these in the left fork arm battery compartment. But this does not really help. One useful tip posted to the RCX400 forum is to purchase fishing weights and put these in the left fork arm battery compartment. These are quite reasonably priced for their weight, and do help. Ball bearings offer another option for weight balance. You can also fit one or more standard counterweights to the fork arm with a wide strip of Velcro. Do wrap the counterweight in Velcro or similar material to prevent the counterweight from spoiling the surface of the fork arm. Also, consider strapping an ankle weight to the fork arm (see Fig. 4.4), to the left fork arm handle, making a reasonably effective, though hardly aesthetic, appendage. It would be great to see something purposely designed for this balancing problem. If and when total balance has been achieved, you should find it quite easy to rebalance next time you change the configuration of your 'scope. Your 'scope may continue to have a westward tendency, but not as badly.

Precise Polar Alignment

If you are a visual observer, your telescope is likely to work well even if your polar axis is not exactly aligned on the north celestial pole. High-end telescopes have many built-in refinements, such as automatic true north and level identification that largely compensate for nominal inaccuracies in polar alignment, allowing the built-in computer to continue to place selected targets well within the standard eyepiece field of view. For most imaging applications, however, particularly those involving auto guiding, a good polar alignment is *essential*. It is also fairly straightforward, either with or without software help.

Visual and CCD Monitoring of Polar Alignment Adjustments

You must first examine the effects of the telescope's offset from the north celestial pole. There are two ways to monitor and correct this. The process can be attempted to some extent and with some difficulty using the supplied 26-mm eyepiece, but at least when performed visually, it is better to use an illuminated reticle. This is a fairly high-power eyepiece fitted with cross hairs, sometimes in the form of a small graticule that is illuminated and is ideal for locating a target star within the small central square. You will need to monitor small changes in the vertical position of a star over a period of a few minutes, a judgment made difficult (if not impossible) without a good reference point within the image field. A reticle makes this considerably easier and more accurate.

Alternatively, far less tiring and more efficient, a CCD camera can be used. If you do jump straight into polar alignment adjustment using a CCD camera, it is *not* essential to know the exact east to west orientation of your image – other than to orientate it so that the longest axis is approximately east to west. During the early stages of setting up your telescope, you may not yet have even "solved" a CCD image – a process performed by an image star analysis program that gives you all the useful details about your field of view, focal length, and camera orientation. This is dealt with in detail in Chap. 5. At this stage, let us assume that you are using the visual method, though we have included other comments where useful. For greatest sensitivity, use the telescope at its native focal length – F10 or F8, depending on your 'scope.

Demonstrating East to West, North to South Misalignment Errors

In almost all cases, if you performed an approximate alignment of the telescope's polar axis using the original solar shadow to obtain due south (as explained earlier), the telescope's actual polar axis is likely to be within a very few degrees (possibly just 1 or 2) of correct alignment. The magnitude of these errors can be demonstrated fairly readily. Locate a reasonably bright star (for example, magnitude 5–7) close to your southern meridian and – for this quick test – near the celestial equator. Center the star within a reticle eyepiece (or with a CCD camera) and just watch. Chances are good that within a few minutes the image will be seen to be drifting either up or down within the field of view. This drift is caused by the polar axis not being correctly aligned in azimuth on the meridian. The top end of the polar axis is likely to be either too far west or too far east of the north celestial pole. You probably know that Polaris is about 43 arc min distance from the NCP. The rate of drift indicates the magnitude of the error.

As an early guide, it is best to aim to reduce this drift to be negligible during a 10-min monitoring. Remember, this is a quick demonstration; try pausing for a moment before you continue with the actual adjustment. Be aware that during these few minutes of testing you may well see the star making small excursions to the east and west around the central point. These slow oscillations are caused by the periodic error in your mount. This error can later be accurately measured and largely corrected (see Chap. 6).

Next, locate a star that is to the east (around 90° azimuth) or west (around 270° azimuth); ideally you should position the fork arms to be one over the other in order to minimize additional errors due to drift from the azimuth error. To avoid the worst effects of refraction, use a moderate altitude (say 30°); repeat the monitoring test (centering a star within the eyepiece or CCD image) and watch. Again the star is likely to drift either upwards or downwards, this time due to the offsetting of the polar axis either above or below the NCP. As before, aim to adjust the axis until you obtain no discernable drift over 10 min.

Manual Adjustment of the Polar Axis

For the precise adjustment of the telescope's polar axis, you are going to rely on the features offered by your equatorial wedge. This is where the cost of the wedge may largely determine how easy it is to accomplish. Wedges made by specialist engineers are far more likely to offer easy adjustment features. When you come to make this adjustment (remember that this section is merely demonstrating the effects of misalignment), you have to release the azimuth adjustment bolts that otherwise lock the 'scope base to the wedge. These are normally retightened at the end of any adjustments. Releasing them does not usually produce any sudden movement of the mount, so you should be able to accomplish this even while a suitable star is visible in the eyepiece or displayed on the screen.

In contrast, when you later release the elevation bolts on the wedge, this allows the full weight of the 'scope to be taken by the latitude adjustment bolt that is about to be turned clockwise or anticlockwise to raise or lower the 'scope on the support plate, so a possibly disconcerting but small move is likely to occur when the tightening bolts are released. Be prepared!

Azimuth Adjustment

Begin with the easier adjustment, the due south position. For greatest image sensitivity, use the longest focal length that your telescope offers without reducers. Remove any focal reducer before commencing imaging – usually at f10. Position the telescope on a suitable star near the celestial equator as described earlier, perhaps in the magnitude range 5–7, and insure that a good focus is obtained. Using the 'scope's keypad, adjust the position of the star, so that it is near the center of the field of view. If you are using a CCD camera, your software may offer a central cross to mark this position. Using *focus* mode, you can start short exposures – for example 0.5–1 s – of the star as you monitor the drift.

Whether you are working visually or with a CCD camera, make a note of the time at which you start monitoring the star's drift away from the eyepiece cross hair or the image center. Unless the azimuth alignment is really bad, the drift should not be excessive and could perhaps take between 2 and 5 min to drift up or down by a few minutes of arc. The aim here is to time the movement made by the drift across an identified (and repeatable) section of the field of view, or to get an actual number for the amount of movement of the star across a portion of the screen. Although useful, you do not need to know the actual screen size. For initial visual checks, you can use a ruler laid across the screen.

The direction of movement is dependent on the east to west offset of the telescope's polar axis from the true pole. Make your first adjustment by turning one of the azimuth knobs on the equatorial wedge visually by a known amount – for example 45° (one eighth of a complete turn). Make this adjustment carefully while monitoring the star on the image (whether eyepiece or screen). As you turn the knob, the star should move across the field, unless there is wedge backlash in this direction. Make a definite adjustment – for example half the field of view – and then, using the telescope's manual controls (RA and dec), reposition the star back at the center of the image. You are now ready to repeat the process of timing and monitoring the star's drift.

After the new time just noted, the star is likely to have drifted by a different amount – either more or less. It may possibly have even reversed direction (that is, gone through the optimum setting and beyond), but this is unlikely at this stage. Importantly, you should see a difference, and hopefully it will be a smaller drift, indicating that you moved the base in the correct direction. When this happens, you have demonstrated that you are improving the alignment!

You can now make a further similar or smaller adjustment to the same azimuth knob. If the drift is greater than in the first measurement, then the knob would have probably been turned in the wrong direction. Therefore, either turn the knob in the opposite direction – allowing for backlash – or turn the knob on the other end of the axis. It is just possible that by chance your mount was set very close to the correct azimuth, in which case a significant adjustment could cause an overshoot. However, in most cases do not expect that to happen.

Reiterative Adjustments

By this process, you can reduce the vertical drift of the star to a minimum amount. As you approach the optimum position, make smaller adjustments in order to try to avoid going through the optimum position. Some published notes describing this process suggest that if the star does not move significantly during a 5-min period, then that should suffice. However, a higher expectation should be adopted. When setting up a mount, you want to see a 10-min period elapse with virtually no drift. It may take an hour or so to achieve this the first time, but you will be able to do it much faster later. However, the polar alignment is usually adjusted only once. When the required minimal drift rate has been achieved, remember to tighten up the azimuth bolts.

Adjusting the Elevation

The process here involves essentially the same process of repeated adjustment, but it also involves releasing the elevation support bolts, so that the whole 'scope is resting on the adjustment rod. First, move the telescope to the east or west, depending on which is the clearer view. You should be looking for a star somewhere around due east or west and at about 30° elevation. Find a star that is not too dim in the eyepiece, or can be clearly seen in a 1-s camera exposure. You are then ready to adjust the mount.

It is highly likely that the star will judder somewhat when the bolts are released, so do not be too surprised. Reposition the star, or locate a new one in the center of the image. You are now ready to start the process of noting the time and monitoring the drift. The process is similar to the azimuth adjustment except that you will be raising or lowering the 'scope's polar axis and leaving the azimuth controls alone. Mechanically speaking, it is much easier to *lower* the 'scope by slightly turning the elevation adjustment bolt to lower the elevation. This is why it was originally suggested that you leave the initial alignment position on the high elevation side. Hopefully you did this!

Try to make the first adjustment carefully and slowly to avoid excessive judder of the 'scope. Once more, you want to know the time that the drift was monitored for and the amount of drift that occurred. Note down whether the drift was up or down, and then make a second adjustment. As before, relocate the star – not necessarily the same one – in the center by using the handset E to W keys and resume your drift measurement.

If luck is with you and you followed the hint about leaving the polar axis slightly high during the initial setting up of the wedge, you should find that lowering the 'scope's axis improves the tracking. The second drift period should therefore see a reduction in the drift amount. You can then make a further adjustment, again lowering the 'scope's elevation slightly and again noting the start time before repeating the drift measurement. As with the azimuth drift measurement, you should be able to minimize the drift such that there is no discernable drift over a period of perhaps 10 min. Remember that if you lower the axis too much, you are risking going through the true NCP, after which you would then see the drift start to occur in the opposite direction. Although you can correct this by subsequently raising the polar axis, this requires much greater effort.

Software Aid in Polar Alignment

The previous, long established method describes the trial and error approach to adjusting polar alignment. It does take time, but accurate alignment can be the end result. The program *PEMPro* was originally written to analyze and derive corrections for periodic error and is described in detail in Chap. 6. This program was later enhanced to encompass polar alignment and telescope backlash adjustment.

Instructions on how to install this software is given in Chap. 6 along with the use of the Polar Alignment Wizard. Following is a brief look at the principles.

PEMPro adopts a more scientific approach than using the drift method described earlier, and could enable a faster achievement of optimum alignment, requiring only that your 'scope be within about 10° of the true pole. *PEMPro's* polar alignment feature works by calculating the amount of movement required to be made, as indicated by the drift direction and the amount of movement of the monitored star in a calibrated field. This drift normally becomes stable after about 3 min or so, so each measurement may be completed within about 5–8 min. Additionally, there may be less likelihood of your adjusting the 'scope in the wrong direction when the instruction actually says *Adjust counter clockwise (west) by 4 min!* However, polar alignment should not be rushed.

A very attractive feature of this (polar alignment) module is the process guiding you toward the adjustment. You are able to select a star and have *PEMPro* show you the direction in which it should be moved during this adjustment. The whole process is repeated for the elevation adjustment.

If you intend doing purely visual observing with your telescope (and why not?) many of the refinements discussed in this book are of only minor concern. Do remember, though, that both the LX200GPS and LX400 are designed for excellent imaging, and that purely visual observing does not really require the extras provided on these top-end 'scopes. The LX200GPS series is more than adequate for visual work.

Next Stage

You have reached the point where some software is essential for progressing to the next stage of your endeavors to get the best from your telescope. Some of it is free, some not. Regardless of whether you move on, your telescope should now be more than ready to delight you and your friends with its excellent images and precision target-finding capabilities. However, with a little more equipment and software, you can make a giant leap.

Chapter 5

Essential Software for Basic Operations

We have described the process of setting up the LX200GPS and LX400 telescopes and using some of the built-in facilities – including *Smart Mount* and *Smart Drive* – to raise the telescope's operational efficiency to a reasonable level. There is further to go with this process, but before that, new software is required. A significant improvement to the tracking performance of a LX200GPS/LX400 series scope (and many others) can be made by measuring and correcting the telescope's periodic error, and that will require specific, advanced software; this is dealt with in Chap. 6. But before it can be done, you need to install telescope/camera control software. This chapter deals with at least some of the software that is necessary for attaining a higher level of telescope performance.

It would be completely remiss of us to not mention the use of computers such as Macintosh and other software platforms such as Linux. There are numerous astronomers who routinely use Macintosh computers for a range of astronomical applications. Some serious programs such as *AstroPlanner* run on all three platforms – Mac, PC, and Linux.

Supplied Control Software

Strictly speaking, some of the improvements to the scope's performance referred to in Chap. 2 apply mostly when the scope is used visually; the whole dynamics of the telescope changes when you fit an imaging system or a guide telescope. Any significant modification to the optics or the addition of hardware affects the weight distribution, and this can dramatically change the balance of the telescope. The telescope should therefore be rebalanced after any change in weight distribution.

L. Harris, *So You Want a Meade LX Telescope!*, Patrick Moore's Practical Astronomy Series,
DOI 10.1007/978-1-4419-1775-1_5, © Springer Science+Business Media, LLC 2010

All of the operations so far discussed can be performed using the Meade-supplied *AutoStar Suite* software referred to in Chap. 2, so no additional expense, up to now, has been required. This is about to change!

Scopes such as the LX200GPS and LX400 series are ideal for taking long-exposure CCD images, for the reasons discussed in earlier chapters. However, the change to imaging operations requires new facilities, including suitable software for telescope control and even automation, and additional equipment to ensure that the telescope is made ready for such operations.

There are a number of ways to control your CCD camera and your telescope. Your camera will almost certainly have been supplied with a basic program as a minimum; for example, the *Starlight Xpress* and Santa Barbara Instruments Group series of cameras arrive with control software. As discussed previously, it is strongly recommended that before installing more advanced software you ensure that your telescope and camera are fully controllable with the software that came with your telescope and camera.

Available programs vary in capabilities; they may merely allow you to expose and download an image, or possibly a sequence of images, but may also offer some limited image processing as well. Some may even offer a basic planetarium program. Several third-party programs include telescope control, and your telescope package may also include such a program. Software that combines both camera and telescope control features is more rare (see also Chap. 8). One such program is *MaxIm DL*. The comprehensive nature of its modules is appealing. It is a powerful program and very useful because of its combination of telescope control, scripting capability, camera control, and image processing options. *AstroArt* provides for image download and processing, while *MaximDL* additionally includes a telescope control module. Careful study of the astronomy forums can provide references to other software that may have potential for full control. *Carte du Ciel* (*Map of the Sky*) is also a potential contender and free! The term 'essential software' can be applied to your chosen program(s) if they can control both telescope and camera.

The ASCOM Initiative

Back around 1998, DC3 Dreams (also known as Bob Denny) developed the *Astronomers' Control Panel* (*ACP1*), a Windows program that gave access via computer to all of the capabilities of the Meade LX200 Classic. DC3 Dreams concentrated on an open Windows scripting interface that included a Telescope object. This resulted in other programs being able to use this for high-level access control for the LX200. During program development, it became clear that if *ACP1* could also control the CCD camera, it could be used for observatory automation. DC3 Dreams contacted Diffraction Limited, and they jointly conceived an open Windows scripting interface for Diffraction's *MaxIm DL/CCD* program. It became possible to develop Windows scripts that controlled both the telescope and CCD camera using *ACP* and *MaxIm DL*.

At that time, there were few astronomy programs that could control hardware (telescopes and cameras), and all required individual drivers for each supported device. There were other software writers with the same interest in developing compatible astronomy programs to match telescope hardware, so the ASCOM (Astronomy Common Object Model) Initiative was born. Support came from Imaginova (*Starry Night* software) that provided a development grant. Eventually, other writers joined the initiative; this led to an article in *Sky & Telescope*.

In mid 2001, the first *ASCOM* platform was launched. *Version 2* was released in 2002 and included a focuser interface standard from Larry Baker and Steve Brady. There were subsequent releases of *Version 2*, each adding more drivers. In October 2003, *Version 3* was released, containing a new dome standard, enabling robotic operation of domes in synchronization with telescopes. *ASCOM Platform 4* was released in December 2004 containing a new Telescope V2 interface specification, together with many updates. There was then a long period of stability, during which time many driver updates were released.

The latest platform – *Version 5* – was released in 2008 and was a significant departure from previous versions. The numerous drivers and their updates had become cumbersome, so the new platform excludes drivers; these are selected for separate download and installation as required. A change from Visual Basic to the new Microsoft.NET family of languages and tools was also made.

The mission to develop software drivers that are independent of vendors and computer language was a good omen for the future. A software (or device) driver is actually a computer program associated with a specific device – such as a printer or telescope – that allows another computer program to interact with it. Typically, the controlling program could be *MaxIm DL.* This permits wide equipment and software flexibility of great benefit to astronomers; any device for which an *ASCOM* driver is available can be interfaced within all *ASCOM*-compatible software. For full details about *ASCOM* you can visit the Web site (see below) and read the extensive information about its operation. For our purposes, there is much to be gained by installing the *ASCOM* platform because this enables *ASCOM* telescope drivers to be used by the software, including *MaxIm DL.* These drivers invariably include more features than the basic vendor drivers, as well as permit *ASCOM*-based software to interact. You will find that advanced programs, such as *MaxIm DL, PEMPro,* and several others are all *ASCOM* compliant (http://ascom-standards.org/).

The *ASCOM* platform – Version 5 – can be downloaded from the *ASCOM* site. Subsequently, load the driver for your specific telescope. There is also an *ASCOM* hub called POTH (Plain Old Telescope Hub), enabling more than one application to access your telescope. POTH is included with the *ASCOM* platform. You might wish to use a hub, depending on your planetarium program, telescope, and camera control programs.

The features described here will largely apply to any program capable of controlling telescopes. For many years, control signals were transmitted along an RS232 (serial) cable, and indeed this option largely remains. It is also possible to have a USB option, and this has the benefit of also permitting camera image data to be configured to use the same cable. For example, an SXV-H9C camera and telescope can be connected via the telescope's USB port to an observatory computer's USB ports. Prior to routine use, the necessary software camera drivers must be installed using the manufacturer's instructions and software. It remains essential to confirm that the supplied software can connect with your telescope and camera *before* trying to connect to the *MaxIm* CCD camera module. This helps to insure that the drivers are correctly installed.

Some Meade telescopes are provided with a CD containing Meade drivers, together with instructions for their installation. The LX200GPS and LX400 are normally supplied this way. From time to time, new updates may be issued, and these can be downloaded and installed. Similarly, firmware updates are occasionally issued and should be installed when available. Announcements are invariably made in the appropriate news forums, as also are early comments on any new bugs that they might include! Generally allow a few days before installing new drivers in case any early users find a significant problem. It is always possible to revert to earlier driver versions, so you need not be overly concerned about an early installation.

ASCOM offers a range of drivers for telescope accessories, including dome and roof drivers for several available controllers. Focuser drivers, a rotator, and of course a range of telescope drivers are all available for free.

Initial Connection to your Hardware

For the first connection of your computer to your telescope, it is suggested to use the native Meade-supplied software driver and not connect via *ASCOM*. The *ASCOM* driver requires a separate configuration, and although this is not difficult – in fact, it is more versatile than the native driver – as discussed, it still makes sense to verify that the basic configuration works perfectly. If any problems are discovered, these can be attended to before the more advanced drivers are loaded. For serious astronomy, you are most likely to want to use *ASCOM*, after the basic supplied software has confirmed that all is working.

Using *MaxIm DL*

Below is a recounting of the author's own experience using *MaxIm DL* for telescope and camera control.

Telescopes: A variety of telescope makes and types work well with *MaxIm DL*. This is because it relies on *ASCOM* drivers for telescope control and, like other modern astronomy software, contains no control code for specific telescopes at all. Where the scope is capable, it can be automatically driven to any object in *MaxIm DL*'s database. This includes all the Messier objects, a large number of NGC, IC, and other classification objects, and many of the brighter stars. A star or an object can be centered and synchronized to the scope, generally improving subsequent pointing accuracy, at least over the region concerned. Scripts can be prepared to drive the scope sequentially from target to target. In a later chapter we shall describe how this is applies using *ACP* (Observatory Control Software) to prepare asteroid measurement runs as well as sequentially targeted long exposures.

Cameras: *MaxIm DL* caters for a variety of CCD cameras. On its own, the module can operate a filter and other devices and also produce a variable-exposure timed sequence of images. A guide telescope with separate guide camera can be connected, calibrated, and used to improve guiding. Adaptive optics units such as SBIG and the SX-AO module can be operated. The usual dark frame and bias frame sequences can also be scheduled using the module (except in some circumstances – see Chap. 8).

Image processing: So many image processing features are included that I am not listing them here. Refer to Cyanogen's Web site shown in the listing. The serious astronomer can complete a wide range of projects from supernova hunting (aided by an included module), variable star studies (using the photometry tool), and deep sky imaging (using image stacking features).

Operations with MaxIm DL

At this stage you have confirmed that the scope can be connected to your computer running *Autostar* and that the camera can be connected to its native software, all proving that the necessary drivers are correctly loaded. The next thing to test is the *ASCOM* software that is now installed for the telescope. A cable (USB or RS232 depending on your scope and facilities) connects your computer to your camera. The LX400 scope can deliver telescope commands and camera data down the output USB port utilizing the scope's onboard USB hub. Having ensured that all the software has been installed, you can activate the observatory module in *MaxIm DL*.

The opening display is the setup tab – see Fig. 5.1 – showing telescope, dome, and focus connections. Selecting "choose" under options provides the drop-down

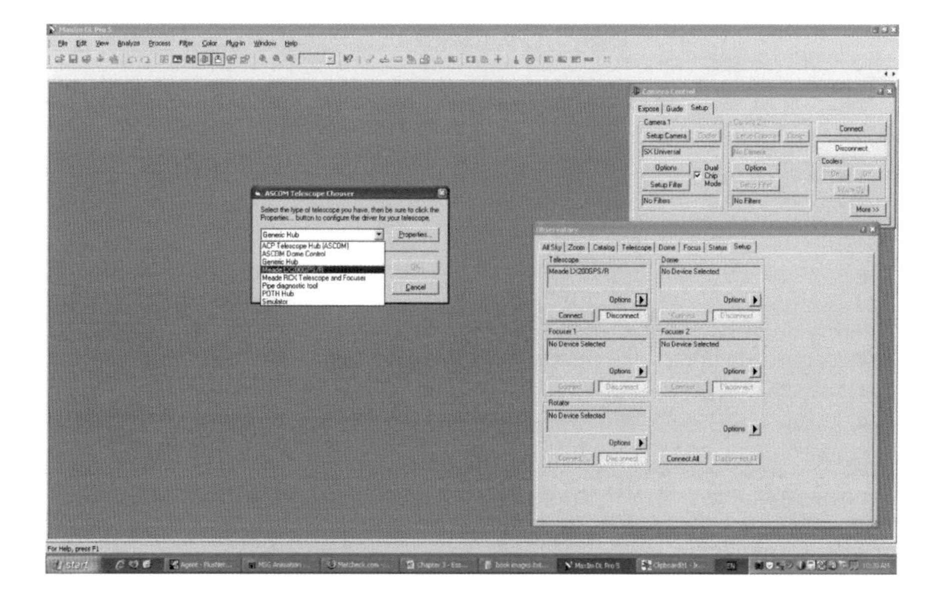

Fig. 5.1 ASCOM options for telescope control using setup tab on observatory control module

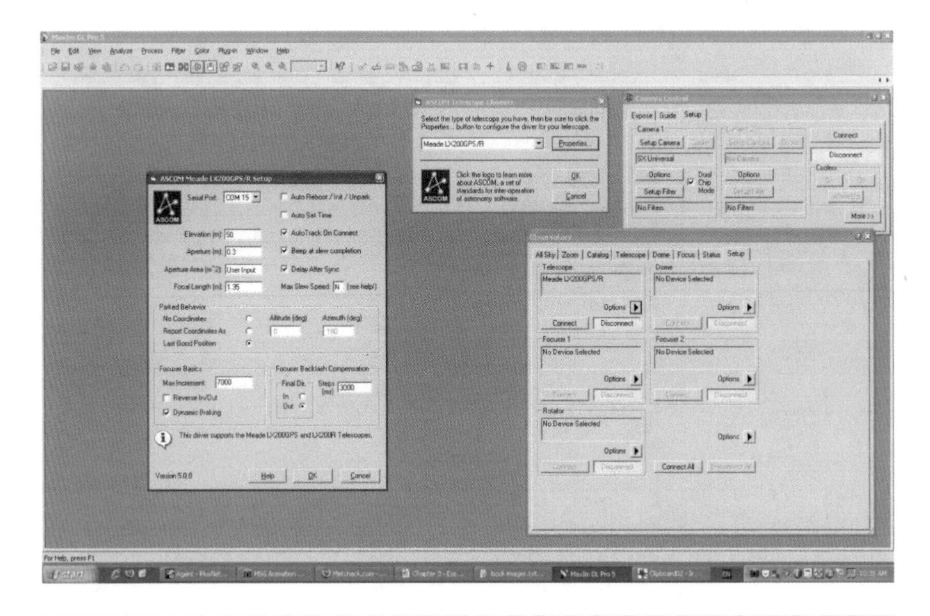

Fig. 5.2 ASCOM properties for telescope control

list of telescope drivers previously installed. After selecting your scope driver (*LX200GPS/R* in Fig. 5.1), you can select "properties" and then enter the options for your driver's preferred state (see Fig. 5.2). Unless you already know which com port your scope is attached to, use the Windows *control panel, system, hardware, device manager* to identify the com port used by the telescope (and/or camera in later trials). Figure 5.2 shows the COM port set to 15, but normally you would use com ports 4 (for telescope and camera) and 5 (for additional optical device) (see Chap. 8 about adaptive optics).

Study the help features if you are uncertain about any of these entries. Have a look at the properties tab (see Fig. 5.2) where the com port entry is shown; this shows chosen settings, including the important *Max slew speed* at N. This latter setting ensures that the scope's own setting of a low slew speed (set within the handbox) is honored and not overwritten. Click "OK" as necessary, and then "connect," and after a few seconds pause, you should achieve a direct connection between your computer and the scope. If this fails (which it should not), you will need to recheck the previous entries.

A closer look at Fig. 5.1 shows the various telescope drivers available after installing the *ASCOM* platform and any required drivers downloaded from the *ASCOM* Web site. The flexibility of *ASCOM* can be appreciated. A number of software writers and a few telescope manufacturers have produced *ASCOM*-compliant drivers that, when installed, will appear in this section. This is where you select the driver most appropriate to your requirements. You can also test specific drivers when they become available. Considerable telescope control is available using *MaxIm DL's* telescope module, which is described later in this chapter. *ACP* Observatory Control Software can help with special projects and is covered in Chap. 8 – Advanced Software.

Camera Control

Most recent CCD cameras include USB ports for control and data access, so these can connect by USB cable directly to the computer. The LX400 scope includes a USB hub in the back of the Optical Tube Assembly that provides three USB connectors – see Fig. 2.6. This means that only one USB cable is required from the scope to the computer (except in very cold weather).

Note comments in Chap. 2 under LX200GPS and LX400 connector panels regarding USB hub stability. With all cables connected and their drivers loaded, the *MaxIm DL* CCD camera control *setup* tab can be selected. The *Setup Camera* option can be selected and set to your particular model. With the SXV Universal driver, a further menu is available for more settings – see Fig. 5.3. This is where the presence of an adaptive optics unit would be registered, in the form of identifying its COM port – "*Select AO*." As mentioned earlier, this com port number can be best identified using Windows *device manager*.

If the plethora of COM port numbers seems confusing you can identify the new one (e.g., the new device) by simply watching the number defined on the *device*

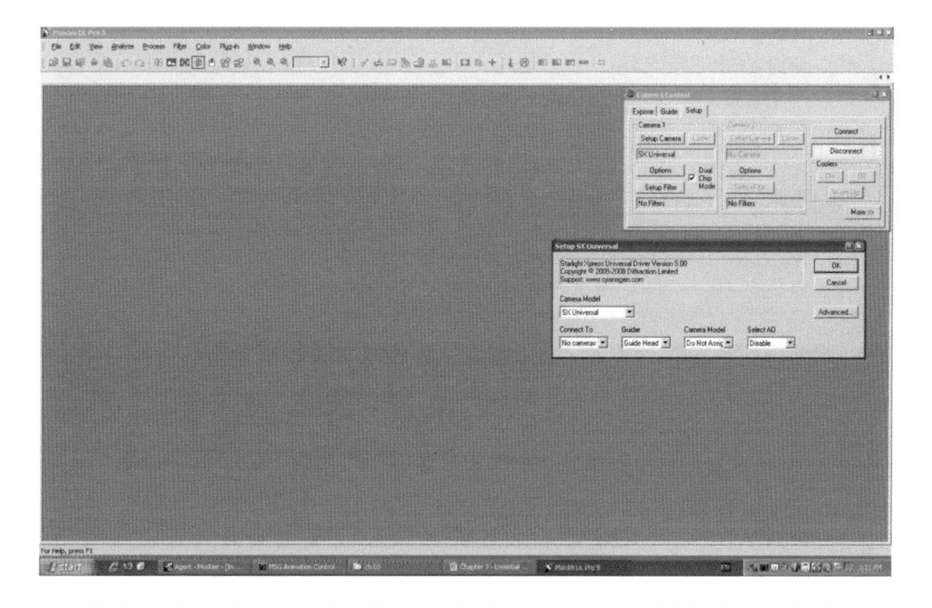

Fig. 5.3 Connecting the camera to MaxIm's camera module using the SXV Universal driver

manager when you plug it in. When all settings are checked, the OK button can be pressed, followed by "connect." As long as the original camera connection software showed the presence of the required drivers, there should not be any problem obtaining a connection in *MaxIm DL*. The actual operations that can be carried out with this module will be examined shortly.

Configuring the Scope: Taking Control

You have connected your computer to the telescope and to the camera using the appropriate *ASCOM* driver(s). The next step is to ensure that the scope is properly focused and aligned on the sky. Select a bright star from a planetarium display; use one high in the south.

At this stage you can either use the keypad itself to select the bright star or do it via software. On the keypad press "*star*" and then select from the numerous stars offered by the Meade built-in database. Press "*goto*" and the telescope should slew. If – heaven forbid – you forgot to select the low slew speed (described previously), do it as soon as the telescope stops moving and gives its beep. It is essential to avoid damaging your motors by preventing them from using the high-speed slew. A cautionary note here: If your *ASCOM* telescope driver has not been set to honor the keypad setting, then using the software *goto* will override this and send the telescope driving at full speed. Ensure – as explained earlier in the *ASCOM Properties* section shown in Fig. 5.2 – that the *ASCOM* driver speed is set to N. This setting remains stored by the software.

The same move can be more effectively controlled using the software. The Catalog option in *MaxIm DL* provides complete access to a huge number of stars, Messier objects, NGC and IC objects, and others. The *search ID* option provides much useful information, including the current slew distance. This tab (*Catalog*) offers a range of very useful ways to produce suitable observing targets on the fly – that is, if you have not previously produced an observing list.

Slew completion is marked by a beep from the telescope keypad. In Chap. 2, the need to align the finder scope accurately with the main scope was discussed, suggesting that a bright star be used for this. You should therefore be able to look through the finder and see the star (hopefully) central. If the centering process was completed at that time, the star should now be visible on a downloaded image from the camera.

Use the *focus* option – see Fig. 5.4 – to produce a binned image (see below) of perhaps 1-s exposure and set a delay between exposures of about 1 s. A binned image is one where you have selected to only download the data in a lower resolution format for speed. The delay is to enable the easy use of the stop button! Note that the downloaded image section shows three out-of-focus stars – the result of using a Hartmann mask for precise focusing.

MaxIm DL provides a feature to select a small portion of the full image to reduce the download time and therefore improve the focusing process. Select a rectangular area around the bright out-of-focus star and resume focusing. With the LX200GPS

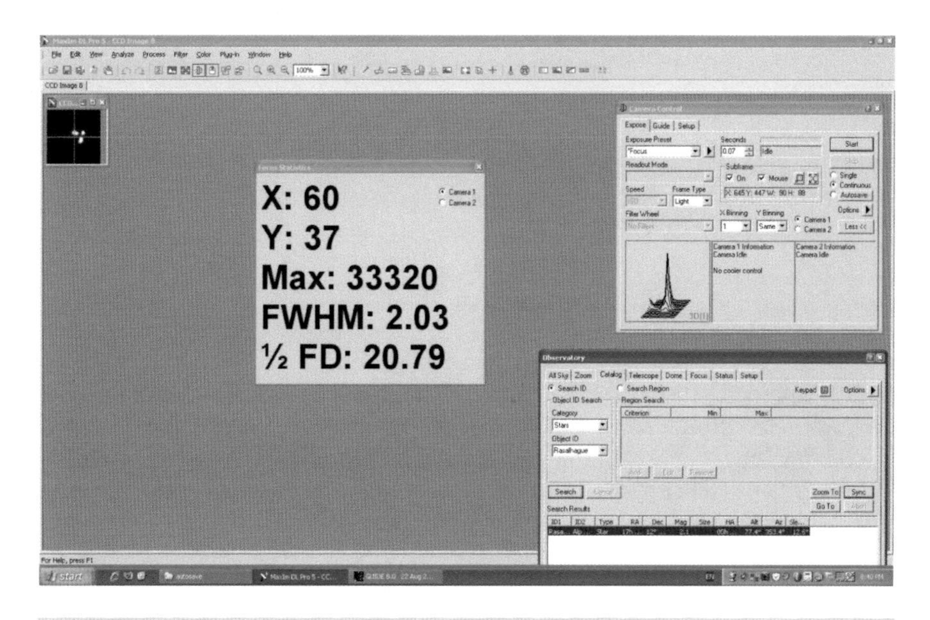

Fig. 5.4 Focusing the main camera using *focus* option and a three-hole Hartmann mask. Focus is achieved when the three images converge to a minimum size

you should first obtain a rough focus using the focus knob. You can then use the microfocuser for final sharp focusing. *MaxIm DL* offers real-time star analysis to display a curve showing the nature of the star's sampled downloaded signal across the full circle. The online *MaxIm DL* help feature explains in detail how to use the FWHM (full width half maximum) indicator to obtain best focus. There really should not be any problems at this stage.

When you have obtained your best focus, stop the imaging and, after selecting bin 1 (the highest image resolution), continue the focusing process until you achieve the best result. Clearly this image will be much fainter than the binned image. You can increase the exposure at this point while ensuring that the sampled data does not saturate. Generally aim for maximum peaks less than 55,000 on a 65,536 maximum (ADU) camera. Again, the FWHM reading is a good guide to perfect focus.

After you have mastered the focus procedure, future sessions are likely to be much faster, especially if you do not change the optical system between sessions. A quick focus prior to an imaging run should suffice for some hours during the night. It is well worth monitoring the focus quality after an hour or two, particularly when the night temperature is dropping rapidly. Except for the LX400, which has a low expansion tube assembly, some small changes of focus may occur during a couple of hours observing. Do not forget to remove the Hartmann mask after focusing – before imaging starts! Also ensure that you use a bright star of high elevation to minimize atmospheric scintillation.

Guide Camera Focusing

At a later time – or now if you have installed a guide camera – you can use the guide camera focus option. *MaxIm DL* provides full control for the guide camera, referred to as camera 2 in *Version 5*. Guide camera focusing is dealt with in detail in the chapter on auto guiding.

MaxIm DL Telescope Header Settings

Before any image analysis can be done by *MaxIm DL* or other comparable program, you have to edit *MaxIm DL's* settings for your scope. Select the *file/settings* option. The menu offers several tabs including *FITS header* and *Site and Optics*. On the *FITS* header tab you can set the *object* and the *instrument* to be added automatically by the software. On the *Site and Optics* tab – see Fig. 5.5 – you should enter the focal length and aperture of your scope. At this stage you know the approximate focal length – if you are using the native scope (F10 or F8) and therefore its focal length in millimeters.

You will probably opt for normal running with a focal reducer (often f6.3), in which case you may need to estimate this focal length. At this point, even an estimate will help the analysis software to identify the region of sky where the image was obtained and its approximate coverage. It becomes important during the different

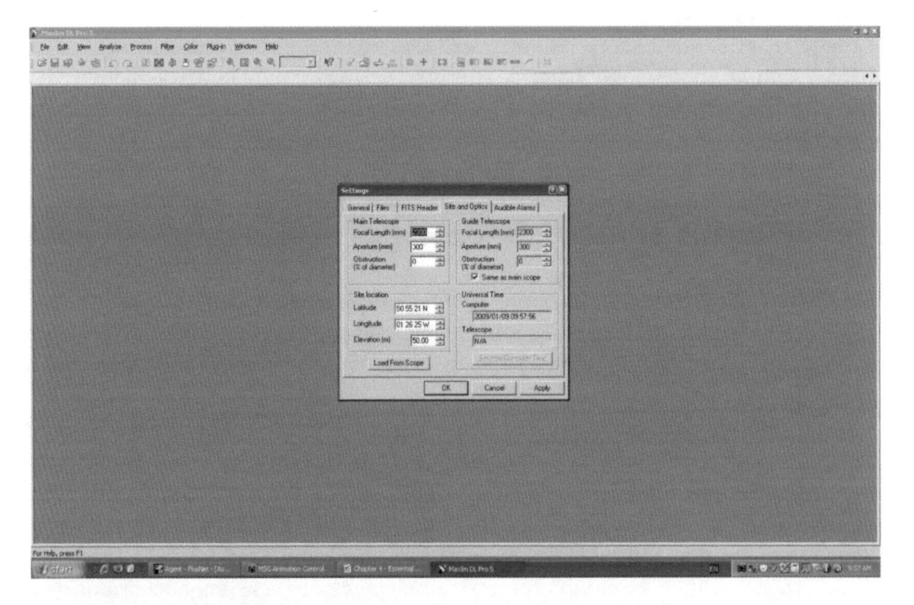

Fig. 5.5 Editing site and optics data in *MaxIm DL*. This data is inserted by *MaxIm DL* into each image's FITS header

observing sessions to remember that changing the optics requires this parameter to be changed. If you forget to change it, the worst that happens is that an inaccurate entry is added to the FITS header of your image. This is by no means a serious error, and it can be fixed; however, try to ensure that it is changed whenever the equivalent focal length is changed. When entries are complete, select *OK*. The next downloaded image will include this data in its FITS header.

Synchronizing the Telescope with the Sky

Both the LX200GPS and LX400 series scopes (and some other high-end scopes) have digital encoders on both declination and right ascension axes. These provide near continuous readouts of the current pointing position of the OTA (telescope tube), which can be read by software connected to the scope. Each downloaded image should have this information automatically added to its FITS header, as just described. This invariably helps image analysis. If all the procedures for polar alignment, GPS time, longitude, and latitude settings have been accurately set (and the built-in GPS receiver can provide this information directly; see Chap. 2), then the scope can be "*parked*" at the end of a session. This is normally due south (on the meridian) and at zero declination. This position is adjustable; it can be changed using the keypad options. When the scope is subsequently powered up after a *parked* close down, the scope remembers where it is pointing. This means that any subsequent *goto* should position the scope close (if not exactly on) to its intended target. Test it by finding a known bright star from your planetarium

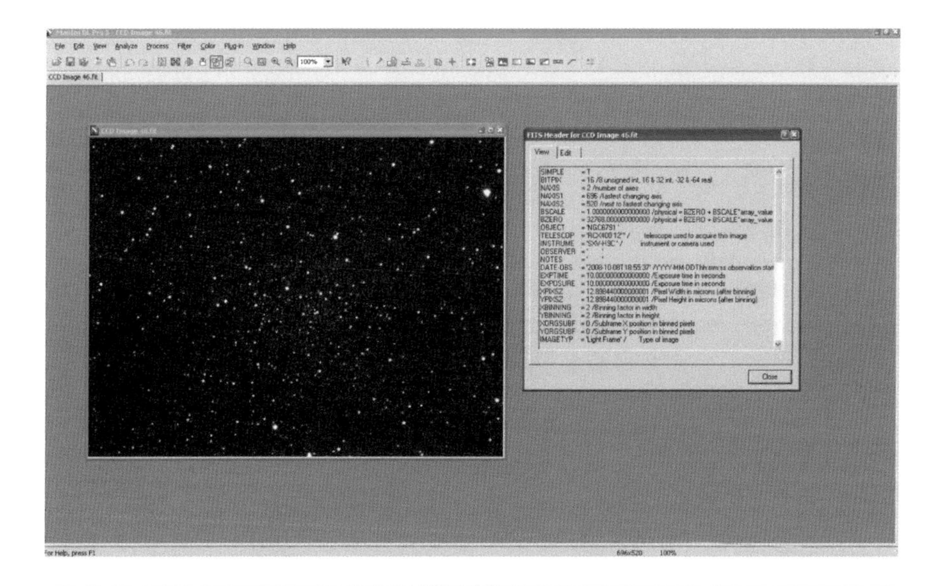

Fig. 5.6 FITS header display for 10-s exposure binned image

program (or the planetarium section of *MaxIm DL*) and select a suitably positioned star. Clicking the *goto* button on *MaxIm DL* should send the scope (at slow slew speed) to the star. The finder should show the star reasonably accurately placed, and it should be within the field of view of the CCD camera.

Having previously focused the camera, expose and download a binned 10-s image. Select the FITS header and check that the settings are correctly present (see Fig. 5.6). This shows the object, telescope, instrument, and observer as entered earlier in *MaxIm DL*'s settings. It also provides pixel size and other data from the software; these data enable the image to be analyzed. That is next on our list.

Solving an Image

This is the process in which you call on the star position analysis software within *MaxIm DL,* called *PinPoint LE* and originally written by Bob Denny of DC Dreams. It looks at the relative positions of each identifiable star in the image and compares them to one of the known databases available to it. The original *PinPoint* is a highly capable program, a description of which we have included in the advanced software chapter because although it is not essential software for controlling the scope, it can be invaluable for use in some of the advanced projects for which these scopes can be employed. *MaxIm DL's PinPoint* is a cut-down version called *PinPoint LE* (limited edition), but even in this form it is highly capable.

Select *PinPoint Astrometry.* Leave the *detection settings* section at their default values, but check the *image parameters.* The approximate FITS center display should be fairly accurate; it is based on the telescope pointing position as provided by the telescope's encoders. The image parameters are based on your input data and whether you are using a binned image.

The catalog setup may need to be entered here if you did not do it previously. Use the *path* option to point to your disk-based collection of star data. You might want to use both the USNO (US Naval Observatory) A2.0 data and the USNO UCAC2 data sets. For general use, the A2.0 data are best, due to their comprehensive coverage. You might also consider using on occasion the UCAC2 data for lower declination target solving. Other data sets are available, including the GSC 1.1 (*Hubble Guide Star Catalog*) data. When the essentials have been checked, you will select *process.* Hopefully the image will be solved, producing a list of basic data about the star field (see Fig. 5.7).

The data displayed by *PinPoint LE*'s calculations includes the following results:

- Matched 295 of 360 images and 850 catalog stars
- Average residual 0.5 arcsec; order 4
- RA 19 h 20 m 57 s, Dec 37 d 47 m 17 s
- Pos Angle +01° 39′, FL 1,490.8 mm, 1.78″/Pixel

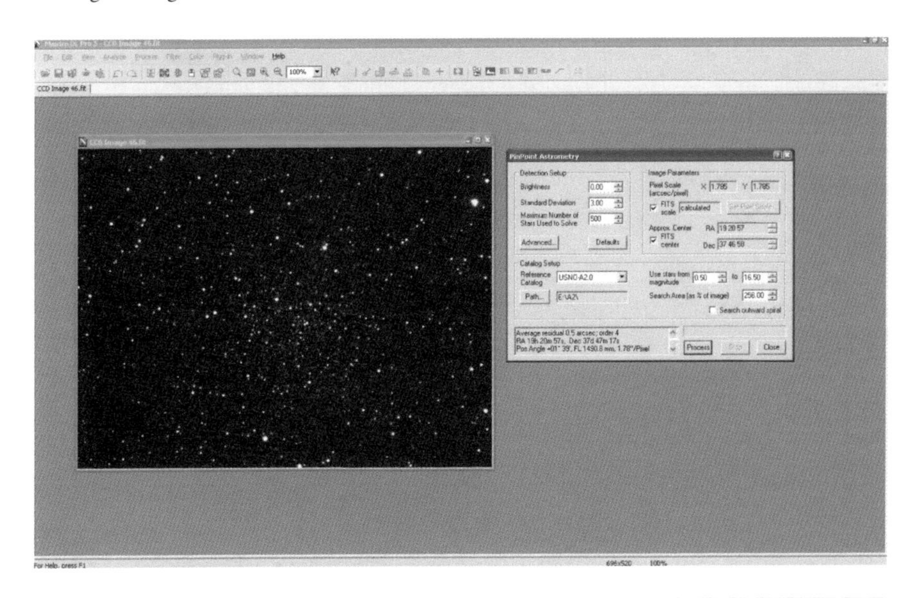

Fig. 5.7 *Pinpoint LE* solved *MaxIm DL* image showing derived data – see text

This tells us that out of 360 stars detected in the image, 295 were used to match with corresponding stars from the catalog (of which 850 were potentially available).

The positions of those image stars were analyzed and found to have average position errors (when compared with the known catalog positions) to within 0.5 arcsec when computed to a fourth order (a high-accuracy measurement).

The coordinates of the center of the image (RA and declination) are then given.

The position angle (measured counter clockwise from north) tells us that the orientation of the image (and therefore camera) is 1° 39′ (i.e., just above zero).

The focal length of the telescope is 1,490.8 mm and the pixel scale is equivalent to 1.78″ per pixel.

Camera Orientation

Pinpoint LE calculates the precise center coordinates of the image, the position angle (PA – degree of rotation) of the star field, the focal length of the telescope, and the pixel scale of the image. It should be noted that the position angle displayed by *Pinpoint LE* is part of the World Coordinate System and must be combined with the plate scale signs to derive the true sky PA. If one of the scales is negative, a *Pinpoint LE* result near zero could actually mean a 180° PA. This data can then be used to correct the FITS data entries if necessary.

At the beginning of any session for which you have changed the optics, use this initial image solve to identify the orientation of the camera and to then adjust it to become close to zero. Have the camera closely aligned with its long axis parallel to the celestial equator. By keeping with a standard orientation, all your images are likely to maintain the same target coverage, which is especially useful with image stacking and blinking (these latter processes are discussed in later chapters).

If the image did not solve (as this process is called), you need to look carefully at the various settings discussed previously. Clearly the telescope's RA and declination must be approximately correct in order for *PinPoint* to know where to start matching the sky! This is the most likely cause of plate-solving failure. The telescope's focal length must be set fairly accurately, though *PinPoint* can cope with moderate errors. The pixel dimensions must be known; these are added by the camera's plug-in driver and will be modified by the binning process if selected.

Another common reason for solve failures is insufficient catalog coverage. If your field of view is of the order of 15 arcmin or less, the Guide Star Catalog may not be able to supply enough catalog stars. If you have a field of view smaller than this, and always when using stars above about 60° declination, you should be using the USNO A2.0 catalog.

Finally, you need to set your exposure interval for initial solving such that enough potential stars to solve can be detected in the image. If *MaxIm DL* reports fewer than 20 stars in the image, increase your exposure interval. Images should contain stars down to at least magnitude 16, if not deeper. With very small fields of view, you may have to go deeper than 16 to obtain enough stars to solve.

Note that on the *PinPoint* window there is an option for star matching to be continued in a spiral search based on the central coordinates. At the early stages of setting up parameters you could leave this on its maximum setting (250%). Once you have shown the telescope, camera, and software to be optimized, you may wish to either reduce this or even switch it off to speed the process.

Synchronization

The *PinPoint* image solution tells you exactly where the scope is pointed; you can transfer this measured (accurate) position to the telescope's encoders by selecting *Sync* on the *Observatory* module's *telescope position* display. This process will often have been arrived at following a *goto* using the *Catalog* option. Adjacent to the option is the *Sync* option, there for this very purpose. This synchronization of the telescope's position will aid future *gotos*, at least in this area of the sky.

To see how effective this process is, select a few more bright stars and perhaps a couple of easily identified Messier objects; use the *goto* option and download a binned image on arrival. The target should be somewhere on the display – possibly even near the center! The slewing process is only accurate if your polar alignment has been carefully carried out. That process was explained in detail in the earlier chapter; software is available to help identify axial movement directions should you wish to perform polar alignment in a more computerized manner. As long as you have balanced the scope, completed the backlash and PE corrections processes, and recently synchronized the scope on a target solution, your telescope should place the target well within the CCD chip after a *goto*.

Field Calibration

On most occasions, following a *goto*, the sampled image is likely to show the target somewhat off center. After *Pinpoint* analyzing the image, selecting sync, waiting for the few seconds prior to the acknowledgement beep, and then selecting goto again, the small residual slew should position the scope very close to the correct position. However, an alternative, rather speedier, method for completing the small position correction is to perform a field calibration.

The best method to achieve this (in *Version 5*) is to solve an image in *PinPoint*. In the *observatory* module, select *telescope* and then *calibrate* (within the *center on image* section). After checking that *type* is set to *auto*, select "use *PinPoint*." The necessary data will be automatically transferred to *MaxIm DL's* settings, and from that point onward you can use the *point telescope here* on the right-click menu. By right clicking on the image at the required new center and selecting *MaxIm DL's point telescope here* function, the telescope can drive to the new position making only small movements. All these features help to increase both the ease and efficiency of telescope/computer use.

Image Processing

Many books have been written about astronomical image processing, some of which explain in great detail how to get the best from your images. For example, we can highly recommend *Creating and Enhancing Digital Astro Images* by Grant Privett (Springer). Many astronomers use advanced image processing software for this. This section includes a basic introduction because it forms part of the process of getting the best results from your telescope.

Longer exposures will generally see deeper into the night sky, though of course there is a limit to the effectiveness of this due to suburban light pollution. Astronomers generally take multiple short- or medium-length exposures – for example, 10-min exposures – and combine these images together in mathematically useful ways to optimize the result. By this method you can build up exposures effectively lasting for hours. But, of course, there is much more to it than merely exposing your camera to the sky. Additional special exposures have to be taken to compensate for built-in electronic defects; there is after all no such thing as a perfect CCD camera. Fortunately, it is not difficult to obtain sets of bias, darks, and flats to get the best out of raw images. The difference between a raw image and one corrected for defects can be considerable, especially in long-exposure images.

Camera Cooling

Because the major noise component added to the final signal is thermal noise, the vast majority of astronomical imaging cameras are cooled to some extent. There is a gradual reduction of this (thermal) noise following switch on. The *information window* shows basic pixel level data derived from the image. Standard deviation is a good measure of overall noise level in the image, and this level can fall from 34.9 just after switch on to about 32 after a few minutes during a test session. If you monitor this once a minute for 10–15 min you should see the figure stabilize. Be aware that this measurement should be done using all covers in place, that is, darks. That way there is no variation due to light changes. With such cameras there may be a longer delay in reaching stability, and in any case temperature variations during a night's session will all have an effect on thermal noise. However, in some cases, including that of several *Starlight Xpress* cameras, the basic thermal noise level is very low. With such cameras there is no need to routinely perform dark subtraction on deep space images; just do bias and (when possible) flat corrections.

Flat Images

The camera chip comprises multiple rows of pixels (picture elements), each of which is sensitive to light. Light falling on the pixels has come through the entire optical system – the telescope. These optics are far from perfect; one effect of some optical systems is that they may not provide a perfectly uniform illuminated image. The field of view is likely to be vignetted, an effect in which the central region of the field is rather better illuminated than the outer section. This effect (vignetting) can be seen fairly readily by exposing an image during late twilight.

Most camera systems can provide short exposures, so when the Sun goes down on a clear sky, an image exposed every few seconds and lasting perhaps 0.1 of a second – see Fig. 5.8 – will start to show a range of shades spread over the field. These are generally called flat images because under carefully achieved conditions they represent the illumination of a flat field across which a known identical illumination is falling. By exposing during a short period of optimum natural twilight illumination you can obtain perhaps 10–20 flat fields. Why so many? Because every single image taken with the camera contains other noise (readout and thermal) that can be significantly eliminated using combinations of other image types (to be discussed). By carefully analyzing multiple identical images, these noise levels can be significantly minimized. The result should be a low noise flat frame image that can be used to calibrate light frame images – those taken of astronomical targets using the same optical system.

What might you expect to see on a sample flat image? Each of the exposures, converted if necessary to remove the Bayer (color) matrix, will show the imperfections, such as dust motes, sitting on various surfaces of the optical system. They will be increasingly out of focus as the surfaces vary from that of the actual chip (where there should be none), becoming doughnuts of increasing diameter as we move further from the focal plane. By the time we reach the reducer's surfaces they are likely to be completely out of focus. The flat therefore gives us an indication of the level of dirt on the surfaces.

A close inspection of Fig. 5.8 reveals three or four fairly large doughnuts but not having a significant effect on the photon counts getting through. This parameter is indicated by examining the image using *MaxIm's information* window. Using the f6.3 reducer shows little vignetting. When a flat exposure has a very short duration there may be little evidence of hot pixels. These have to be dealt with separately.

Bias Images

When all the pixels in the camera are sampled at download, even if everything else is constant, there is an individual offset – called a bias – for every pixel. Fortunately, it is not difficult to compensate for bias offsets. Camera exposure software normally offers a bias exposure setting that will be for an extremely short period of time (ideally of zero length) under dark (aperture covered) conditions. Even this bias reading is itself

Fig. 5.8 LX400 telescope with SXV-H9C camera producing a sky flat by using very short exposure. This flat shows several doughnuts

subject to a readout noise, so again you must take a large number of bias images and combine them mathematically to suppress this noise as much as possible.

Readout noise is one parameter specified for CCD cameras and should be very low. Bias offsets are fairly constant over long periods of time, so one set of bias images can be retained for subsequent use during several months. The SXV-H9C camera has an extremely low thermal noise level anyway, so 10–20 bias images can be processed to provide a very low noise master bias image. A single bias image shows a uniform noise level when displayed across its full range of values (0–65,536). Few artifacts – semirandom groups of nonaverage pixels – are seen.

Dark Images

We are not able to operate our cameras at absolute zero, and at all temperatures above this there will be a disturbance from thermal noise. The amount varies at different cooling stages. The warmer the camera the more thermal noise it will generate. There is also the additional readout and bias noise added to the thermal noise. It is therefore a common practice to largely compensate for thermal noise by subtracting a matching dark frame – an image of the same duration as the light frame but taken in perfect darkness (with the covers on) and ideally at the same temperature. Dark frames naturally include readout and bias noise, but as has been explained, these can be adequately subtracted.

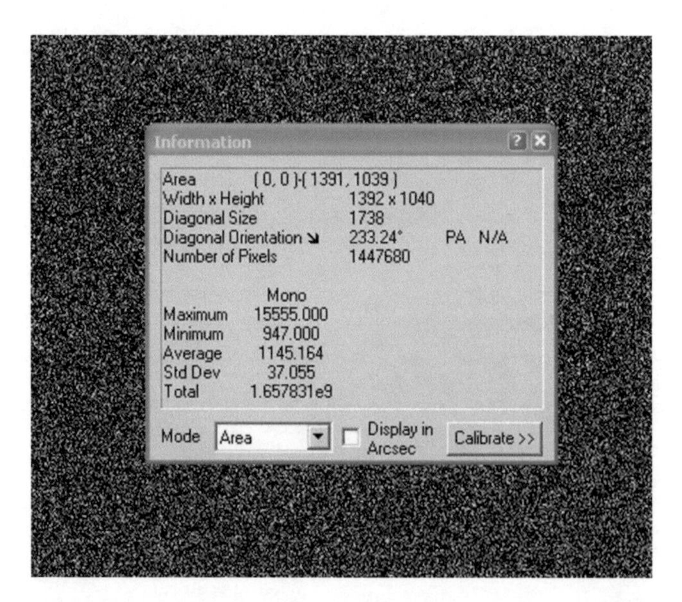

Fig. 5.9 Single 60-s exposure dark frame showing information window

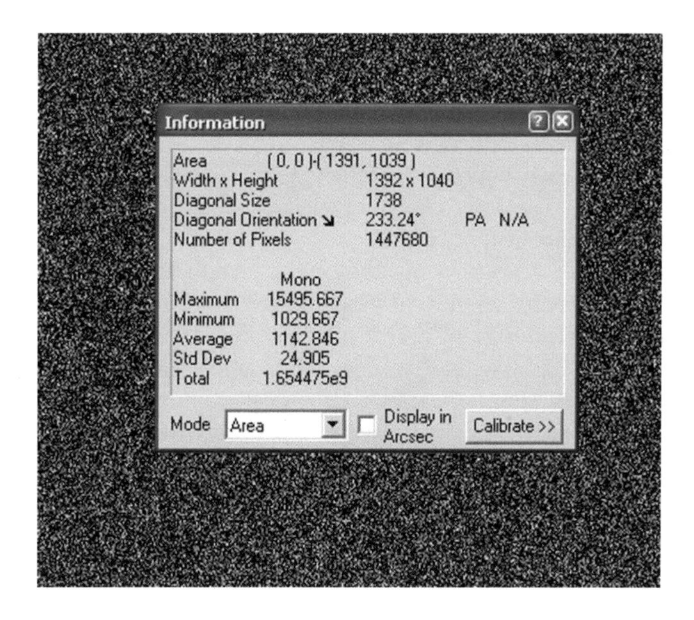

Fig. 5.10 Multiple 60-s dark frames combined to reduce noise level

Significantly, a number of CCD cameras incorporate very low noise chips, so in some cases it is even possible to exclude dark frame, especially when exposures are not longer than a few minutes. By taking a large number of dark frames and mathematically combining them (as with the flats and bias images), the resulting noise level is considerably reduced. In truth, rarely will you need to do dark frame subtraction due to the low thermal noise of the chips, but sometimes you might want to take a set for checking hot pixel monitoring and calibration purposes.

Figure 5.9 shows a single 60-s dark frame displaying a standard deviation (noise level) of about 37 or about 3%. Figure 5.10 shows the result of combining only a few 60-s dark frames into a master frame. The process produces a standard deviation of about 25 down to 2%. This shows the value of taking multiple dark frames in order to provide a master dark frame. A treatise on the subject of image processing will explain the nature of the process, but that is beyond the scope of this book.

Try Some *Gotos*

Having proved that your telescope is now fully controllable by software, it is a good idea to take some sample images. Do not yet use any autoguiding features. (Autoguiding is described in Chap. 7; it enables you to greatly extend the exposure time of individual images.) At this stage, keep exposures short enough to ensure that most of them do not show star trails due to poor tracking. Use a planetarium program (either *AutoStar Suite* or that within *MaxIm DL Version 5*) to identify some suitable targets, and use fairly high elevation, bright Messier objects, such as globular star clusters, for early targets. You could obtain a few series of exposures of the same target for later stacking (adding together), although the next chapter should enable you to extend the exposure time.

When possible, draw up your own observing plan to help maximize your productivity. Without having any idea of what you might do, you can easily find yourself spending valuable observing time just wondering what to do next! You can use *Guide-8* for this because you can select suitable targets and then check their optimum times of highest elevation. They are then listed in time order, after allowing enough time for repositioning the telescope after each series of images.

Earlier in this book we explained about the importance of producing a good PE (period error) correction when you are ready. The detailed process of measuring and correcting your scope's PE is described in Chap. 6. To get your telescope ready for this process you have to have the essential software fully configured, hence the need to have described the telescope software configuration process before the chapter on PE. Now that you have ensured that your telescope/computer/CCD camera combination can reliably take images, you can complete the PE chapter. What follows will test the scope's abilities: to see for how long you can expose before inadequate tracking shows and without any form of autoguiding.

You have now taken almost all the steps necessary to bring your LX200GPS/ LX400 scope toward its peak operational efficiency. The next stage of this process is virtually the final adjustment – minimizing the PE of the scope.

Chapter 6

Software Adjustment of Periodic Error and Polar Alignment

In this chapter we look at the subject of telescope periodic error (PE) and discuss images taken by the author prior to correcting his LX200 Classic telescope's PE and subsequently that of his LX400. Software is used to measure and change the PE, remeasure it, make a second correction, and finally make a third measurement! Toward the end of the chapter, we discuss using the same software to measure and optimize the telescope's polar alignment. Please be aware that the larger MaxMount telescopes (40 cm and above) have different gearing; some of the following discussion applies largely to the sizes from 35 cm downward, although the processes of measurement are the same.

Measuring Periodic Error

Establishing a really good polar alignment and ensuring that your telescope's periodic error is reduced to the minimum possible comprise two very important factors governing the overall performance of your telescope. In Chap. 4, we looked at the visual versions of performing polar alignment by the drift method. An alternative method (and one surely easier to perform) involves using a CCD camera to monitor the same drift. That procedure should be carried out after the telescope has been well balanced, hence its inclusion in that chapter. Drift alignment is a popular method of achieving polar alignment and can lead to an accurate positioning of your telescope's polar axis, enabling you to track well for long periods without excessive declination drift. By the end of Chap. 4, assuming you had successfully balanced your telescope, you should have been able to achieve an adequate polar alignment, good enough for you to be able to proceed with periodic error correction.

Ray Gralak's program suite *PEMPro* (*Periodic Error Management Professional*) is used to measure and correct periodic error, but it can also help with the adjustment of polar alignment in a measured manner (the latter process is a more recent

L. Harris, *So You Want a Meade LX Telescope!*, Patrick Moore's Practical Astronomy Series,
DOI 10.1007/978-1-4419-1775-1_6,

development of *PEMPro*, and a description of it follows later in the chapter). Included here are real data from periodic error measurements on two entirely separate occasions, to help explain what can be done to greatly improve the tracking capabilities of a telescope that is suitably equipped. By "suitably equipped" we are referring to the feature built in to the LX200GPS and LX400 series telescopes, and often available on other higher priced telescopes, that enables the user to measure and permanently modify the periodic error that is native to all scopes.

What Is PE and Why Do You Need to Fix It?

The LX series telescopes (LX200GPS and LX400 series) use a small DC (direct current) motor to drive a small gear reducer train. This reducer train turns the worm (essentially a threaded rod) that then drives the final worm wheel that points the telescope on the RA axis. None of these individual components, singly or in combination, is perfect. Indeed, there is no such thing as a perfect gear train or perfect worm drive. The inevitable imperfections that arise naturally during the mass production of telescope components result in small but visible errors throughout the full cycle of your telescope's drive.

Most of these errors are, by their very nature, periodic. As will be repeated more than once in this book, for routine visual use – that is, using eyepieces – such variations are usually tiny and of little consequence to the observer. The eye naturally follows the target object such that position variations experienced while the target slowly meanders within the eyepiece field of view should hardly be noticed. For the imaging enthusiast, however, it is a very different story.

Some of the most expensive mounts that the amateur can buy already have excellent tracking capabilities and minimal periodic error, though even these can, in principle, be improved using the processes described in this chapter. A glance at the manuals for the LX200GPS and LX400 scopes (these manuals can be downloaded from Meade's Web site) shows that they are potentially capable of having their periodic error reduced to the level of a few arc seconds. One specification suggests 5 arcsec for the LX400. However, before you start the process of measuring PE you must first check that the following basic operations have been done.

Essential Previous Checks

There are three important matters to attend to before you start the improvement process. If these are not first corrected, your actual measurements of PE are liable to be distorted and inconsistent.

1. Balance: There is a separate section on balancing your telescope (see Chap. 4). An unbalanced scope cannot perform at its best because the stresses within the telescope and its mounting are likely to vary depending on where in the sky it is pointing. The scope must therefore first be balanced. It is a common practice to leave the rear

Fig. 6.1 The declination motor on the west fork arm of an LX200GPS seen with cover off (Courtesy of Robert Kessler)

end (whether eyepiece or camera) a little heavier so that there is a slight tendency for the scope to tilt toward that end. This is simply to avoid the declination axis worm from frequently disengaging with the clutch and floating. The RA axis is likely to require an additional weight on the eastern fork arm to help offset the extra weight of the declination motor (see Fig. 6.1) on the west arm.

2. Drive training: This process is designed to minimize the backlash inherent in both drive trains. See Chap. 2 for further details.
3. Polar alignment: Although this need not necessarily be perfect, it significantly helps when the polar axis is within a couple of degrees or less of the true north (or south) celestial pole: Chap. 4 includes details about completing this procedure to a good level of accuracy.

Out-of-the-Box Images

With these essentials completed, you are ready to test and measure the telescope's current performance. Before you start any measurements, first see what you already have. If a short-exposure CCD image is taken using an LX200GPS or LX400 series telescope without any modifications to the drive's periodic error – an out-of-the-box result – you are likely to obtain pictures such as those seen in Figs. 6.2 and 6.3, with a range of 30–300 s exposure. The telescope should be carefully focused before imaging. No autoguiding or any other enhancement was used for these exposures.

The first image – Fig. 6.2 – is a short exposure using the mount's built-in factory PEC (this will be described later). It shows reasonably tight stars because the mount

Fig. 6.2 M15 30-s exposure using nonmodified LX400 running out-of-the box (factory) settings

Fig. 6.3 M13 5-min (300-s) exposure using nonmodified LX400 running out-of-the box (factory) settings. This is a long exposure for an unguided image and therefore shows evidence of tracking errors

has only been tracking for 30 s. You could probably take 120 such exposures sequentially (a total of 60 min) and find that virtually all were this good. You can double the exposure (to 60 s) and find that 60 consecutive 60-s exposures would very likely be fine as well, or mostly good. Doubling to 120 s is likely to produce several problem images with perhaps an occasional good one. The final image – Fig. 6.3 – shows that a limit to exposure length has been crossed. Tracking errors are obvious.

Remember that no two sequential and equal exposures are bound to show exactly the same amount of periodic error, particularly when exposed for longer periods. It is a characteristic of worm drives that drive variations (varying from perfection) are periodic; there may be periods of excellent tracking, broken by short periods of poorer quality tracking due to momentary tight or loose spots on the worm. Significantly, these tight and loose spots are likely to be associated with specific sections of the worm – hence the periodicity effect.

How Long Can You Expose?

With many telescopes, a 60-s exposure is close to the limit of an unguided image, especially with high-resolution cameras – such as the SXV-H9C – where almost every tiny tracking error shows! Much longer than this and many images show increasing amounts of star streaking. Under perfect conditions, which can rarely be realized, the mount may be able to track accurately for a few minutes. Secondary problems, such as atmospheric scintillation or other factors such as misbalancing, soon start to manifest themselves, causing imperfect tracking.

Two-minute exposures really demonstrate the tracking capabilities of a mount. Such exposures often show tracking errors, and by 5 min the incidence of tracking failure is high; this is a long time for an unguided mount. Unguided means that the mount is left under its own normal, unaided tracking capability, in which the mechanical construction of the mount – the accuracy of manufacture of its primary worm gears – defines the basic drive accuracy. These top-end Meade mounts have advanced features to improve their tracking and guiding accuracy.

If you were to take a series of consecutive 5-min exposures, you might find that perhaps one out of several was more or less OK. This would indicate that one or more short sections of the drive teeth or worm were not too bad. However, when you spend valuable time imaging, you do not want five out of six images to be thrown away! Fortunately, you are about to significantly reduce your telescope's PE.

Measuring and Correcting Your PE

Periodic error correction (PEC) is – when available on the telescope – an excellent feature for those wanting to improve the quality of their imaging, and there are at least three methods of doing this:

1. Visual: In this original but now outdated method, you visually monitor a star in a high-powered eyepiece. During a defined period, the telescope's computer

records your manual corrections and the relative times at which they are made. On an LX200 Classic, this required getting a suitable star visible in an eyepiece fitted with an illuminated crosshair (an essential feature) and then waiting for the zero moment to approach (anything up to a complete cycle). From that precise time, you meticulously watched the position of the star and concentrated on keeping it central on the crosshairs using only the W and E keys on the handset. You had to do this for a complete 8-min cycle, at the end of which the corrections were immediately effected and tracking improved.

This is the method that many astronomers used during the early years from around 1992, when permanent PE correction became available on high-end amateur telescopes. The Meade manual includes a description of this process; on all recent versions of *AutoStar II* it is essential to continue the monitoring process for full 24 min – three complete 8-min cycles. This manual process still does work and the results are good. But it involved a high level of visual concentration for a complete cycle of the worm! You could do a refine cycle if you wished, in order to try to further improve the resultant tracking (however, see later comments). At this time CCD cameras were falling in price, but only slowly approaching amateur affordability, and no software had yet been written for PE correction.

2. Autoguider method: If you have set up an autoguiding telescope mounted on the top of your main scope, you can introduce a level of automation for PEC measurement. Although this method is far less tiring than the essentially manual method described above and does not require new software, it does have limitations. All the corrections produced by the autoguider are clearly done "in arrears" (just as are the manual corrections), such that no matter how short the exposure is, there is an inevitable delay before corrections are issued. The relatively short focal length of the guide scope may also significantly reduce the sensitivity of the measurement.
3. Software measurement: This method has to be the winner. Ever since Ray Gralak's *PEMPro* appeared, this author has used it every few months in order to ensure that his scope remains tracking at maximum efficiency. In the journey to (self-assessed) imaging success (within the limitations of suburban life and our atmosphere), the optimization of periodic error has remained high on the list of things to do. Therefore, we will take a detailed look at how this can be achieved. Before we look at *PEMPro*, though, a bit of background information could be useful.

PEC Settings

LX200GPS/LX400 telescopes have three states of periodic error setting: (1) factory PEC activated – the normal setting; (2) PEC zeroed – a deliberate process to be done later by software that deletes any stored corrections from the PEC table; and (3) a user-produced PEC table replacing the factory PEC. Setting 2 (zeroed PEC) would be likely to show the worst tracking, with setting 1 in the middle, and setting 3 – a PEC table optimized for the particular scope – producing the best tracking. In normal operations setting 1 is used. Figures 6.2 and 6.3 were taken using the default factory PEC setting. In this chapter we shall show how to create setting 2 (zeroing your PEC table) on the way to setting 3 (your own personalized PEC table) – the best of all.

It is worth pointing out the following important information: the *AutoStar II* setting "Erase" is now "Restore to Factory." That change happened with the release of *Version 4.0* of the firmware in 2005. In the LX200GPS, "factory restore" really is "erase"; in the LX400, it does restore the original factory PEC data.

The Details

Before reading the next section, please be aware that it is not really necessary to understand the exact mechanical construction of Meade's gearing to measure and correct your telescope's PE. Some details are included here for those who do like to find out what is under the hood! For even greater detail with commentary, visit and join the LX200GPS forum, and check out the reference sites listed here.

The LX200GPS and LX400, with apertures below 16 in. (40 cm), perform PEC operations in exactly the same manner using firmware current as of 2008. Their gear drives are the same; both have 180 teeth on the final gear, and one worm rotation takes 8 min. Both have programming that currently measures three turns of the worm (3 times 8 min) – hence, 24 min – to obtain a more representative measure of the final PEC error. If you had firmware prior to Version 3 on an LX200GPS it would only perform a limited PEC train. In this PEC process we shall describe a complete three-cycle PE measurement of 72 min (3 times 24 min) that applies to scopes up to 14 in. (35 cm). The process of producing a correction curve is identical for both scopes.

Your scope has a PEC table; this is a list of corrections applied to the sidereal drive rate that the mount will make at a number of specific points during its 24-min cycle. The table is stored by the telescope's built-in computer. The worm shaft that drives the scope is indexed by means of a small magnet inserted in the shaft that triggers a pulse at a point defined as the beginning of the worm cycle. This forms the reference point for the start of a PEC cycle. However, to fully characterize the intermediate gear train's whole cycle, as explained, the LX200GPS and LX400 require three complete worm turns. Both telescopes break the worm cycle into 200 segments, so (3 times 200) 600 segments allow for the complete gear train. Thus, one complete cycle, to have a count of 600, takes about 3 times 8 (=24) minutes to complete.

This 600 segment whole represents one complete cycle or data set; to repeatedly examine – and therefore characterize – the worm for more than one full cycle, it is recommended to monitor three complete 600-segment periods.

A small DC motor drives the gear reducer train. This in turn rotates the worm that drives the actual worm wheel that points the telescope on the RA axis. The RA axis drives at the rate of 360° in a sidereal day. (This day is, of course, about 4 min shorter than a solar day.) The LX200GPS/LX400 worm wheel has 180 teeth, and the worm that drives it rotates one complete turn every 8 min. The small gear reduction train has a reduction ratio of 50 times. This is a very slow speed for a DC motor, so it has to be controlled by a digital signal. This signal generates the motor's full torque by the application of full battery power at all speeds. The actual speed is controlled by the duration of the pulses – a style of control called Pulse Width Modulation (PWM).

Consequences of the PEC Table Adjustment

Rapid changes in drive motion are *not* tracked during the 2.4-s periods that comprise the PEC table's highest resolution (600 segments divided by 24 min give 2.4-s intervals), so abrupt bumps or valleys on the worm gear or worm wheel that occur more quickly than 2.4 s may result in an abrupt image shift. If you do training visually, the averaging of new training with stored training (the previous value) has the effect of dominating the older training. In the case of *train* mode the training value is stored, but in the case of *update* mode the training value is averaged with the older value. Consequently, older trainings become less significant, and the most recent training becomes one half of the stored correction. This does mean that recent training is the most heavily weighted and thus the most important. A poor visually prepared update can therefore spoil an otherwise well-trained worm.

A good training session can therefore be spoiled by a final bad training session if using update, so the averaging algorithm used is not the best. Clearly, in this manual (visual) method of training it would have been better to play back the PEC during training and then allow for corrections, which would be small, to be weighted into the values that are stored, thus perfecting the training in a systematic way as you do more and more training.

The author first realized what was going on with this seemingly strange method of *not* applying the previously measured PE corrections during an early update session because he had decided to repeat the visual challenge of doing a further 8 min on a really clear night. It quickly became obvious that the update option was as difficult as the training session, because it simply repeated the *same* significant visual variations. When a complete training cycle has been performed and a PEC curve produced and implemented, regardless of how it was done, you will normally see a significant improvement in performance. However, it is important to remember that although you have trained the complete worm drive, this has been done on only a very small portion of the worm wheel itself. Under some circumstances – such as after manually moving the telescope around in RA – the PEC training may be less effective than before. There is therefore good reason to leave the drive clutches locked so that the trained part of the worm wheel is used at all times. This is usually practical only on a permanently mounted telescope.

The worst of the worm drive's PE irregularities can usually be considerably reduced by PEC training, leaving only residual – lower – rates of deviation. These residual errors should be correctable with subsequent automatic guiding with a CCD imager. By the time one gets to the 1 arcsec guiding condition, other effects such as telescope mechanical instability, wind-induced motion, and atmospheric variations will take over in blurring a star's image.

After taking a look at PEC and how it has been done in the past, you surely want to see how the whole process can now be completed smoothly.

Periodic Error Management by Software

PEMPro, like a number of other astronomically orientated programs, requires a standard software platform – a sort of *Windows-XP* for astronomers called *ASCOM: Astronomy Common Object Model*. (This was discussed in Chap. 5.) This platform is installed by downloading the program from the *ASCOM* Standards Web site. After this installation is complete, you should then install the *Microsoft .Net 2.0 Framework*. All the essential Web address links are provided. Finally, you can download and install *PEMPro* (http://ascom-standards.org).

The *PEMPro* software allows you 60 days of full operation so that you will have plenty of time to try the program thoroughly. There is an extensive help section that covers virtually all topics and questions that you might have. *PEMPro* works with a large number of telescope types and also with many different types of CCD camera, including Web cameras. There are wizards to enable you to proceed smoothly through the various stages of setting up. Install a copy of *PEMPro* after downloading it from the CCDWare site. Note that other CCD imaging softwares, including other equipment optimization softwares are also available on the site (http://www.ccdware.com/).

From the Setup page of *PEMPro*, there are options for you to enter the basic details of your scope. The page shows tabbed options for PEC, Polar Align Wizard, Star Finder, and Backlash options. *PEMPro* also offers the option of connecting to different telescopes with different optical configurations, and you can save each configuration for later use. When remeasuring PE using *PEMPro* on a 30-cm f8 LX400, its natural focal length (f8) for maximum spatial sensitivity is used. This gives *PEMPro* the highest resolution possible – 0.65 arcsec with the author's CCD camera – much smaller than atmospheric scintillation! *PEMPro* offers a number of default settings, such as minimum star brightness, and these may well be found to be adequate. For your trial run allow at least an hour or more of test time. Configuring the software will take several minutes. Familiarity with the operations of *PEMPro* will engender confidence in obtaining good results.

If you are running *PEMPro* for the first time, use the *Mount Quick Configure Wizard*. You will almost certainly be using the *ASCOM* driver, so this box can be checked. The help section is extensive and should be studied before proceeding too far. The next task is to identify the image scale of your CCD (the arcseconds per pixel value). There is a wizard to help with this. You do need to have one of the following programs installed: *Maxim DL, Software Bisque's CCDSoft*, or *AstroArt*. *PEMPro* also natively supports the complete line of Meade DSI cameras, currently including DSI/DSI Pro, DSI II/DSI II Pro, and DSI III/DSI III Pro. For the pictures in this book we are using *Maxim DL* and an SXV-H9C camera. You may already have identified the image scale of your telescope/camera optical system; if you have solved an image (see Chap. 5, which explains this process), one of the displayed parameters is the image scale. Except for PE measurement, you can use an LX400 f8 telescope with a Celestron f6.3 reducer at an effective image scale of about 0.93 arcsec/pixel. It is useful for you to know the image scale in this manner to serve as a check when you run through the *PEMPro* image scale wizard. Please note, though: for the best results (the most sensitive) you should use the telescope's native resolution – no focal reducer! For PE measurement and correction always use or f10 as appropriate as mentioned.

Finding a Suitable Star

You are going to calibrate the telescope and camera combination using *PEMPro*. Let us mention first that *PEMPro* has its own Star Finder section that does the process in a methodical manner. This module performs a spiral search around the telescope's current position and builds a large image. From the results of this you can continue with the next stage.

However, an alternative procedure involves using a short manual search. Start by pointing the telescope due south between 0° and +20° declination and then try to find a star – or better still, a group of bright stars, as will be described shortly. There has been useful comment on the forums, suggesting that equally accurate PE measurements can be obtained using stars at higher declinations, even up to about +60°. This probably works. For your early measurements, use declinations between 0° and +20° as often recommended and as discussed here. After you have validated this procedure for yourself, you can then try higher declinations at a later date.

From this start position, while watching the full CCD field of view using 1-s binned exposures in *focus* mode, slowly sweep westward for up to 10° in RA (that is, no further than to 190° azimuth). You are looking for at least one star of nominal brightness such that you can take 0.1–1 s (0.5 is best) exposures; this is short enough to expose frequently while not so bright that it saturates the sensors. A star seen in binned focus mode at 1-s exposure should be fine for sampling at 0.5 s in full resolution. If, unusually, no suitable star is quickly seen, return to around 182° azimuth and then drive northward slowly for a few degrees until a star is found. This azimuth position allows enough time for an 80+ min measure of PE. You should be able to find a suitable star using this slow sweep method.

The *Setup/Calibration* section can take you through the process of finding a suitable star for monitoring so that you can then do a test run. If you have just followed the notes in the previous paragraph, you should already be there. The next process (see Fig. 6.4) establishes the exact image scale of your camera's field of view – an essential measure. As long as you have correctly configured the software and your telescope, the wizard will help you to produce a field showing several fairly bright star trails. These are produced by *PEMPro* switching off the telescope's drive for a selectable number of seconds and then switching it back on.

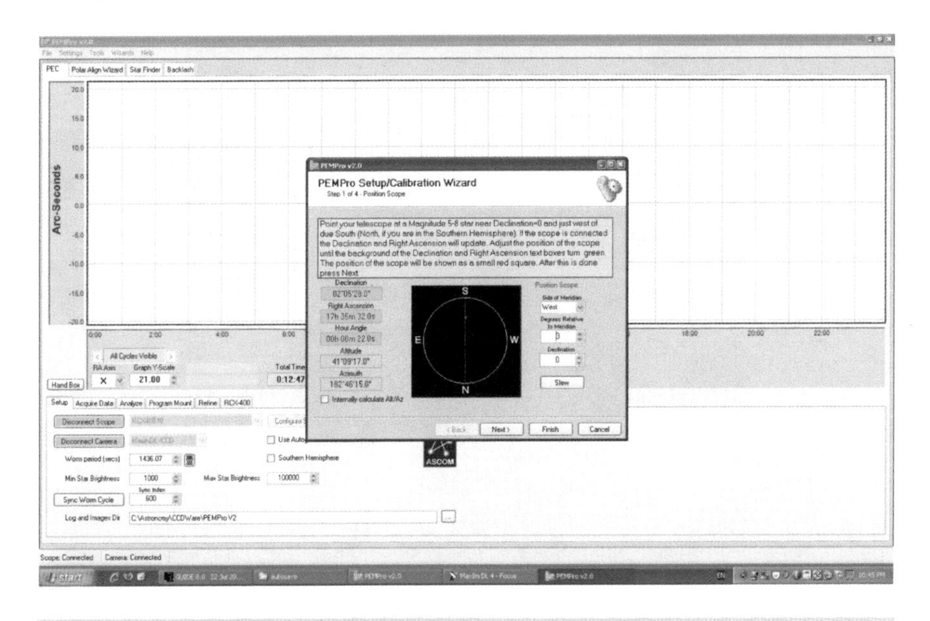

Fig. 6.4 *PEMPro* image scale calibration using the wizard. This is an essential early measurement

Fig. 6.5 *PEMPro* star trail ready for measuring after *PEMPro's* mount drive pause. Clicking on both ends measures the image scale and orientation

Figure 6.5 shows a section of a field with one bright star trail ready for measuring. Ideally try to find a field that includes several bright stars. *PEMPro* should then produce a set of star trails. This allows you to point and click the beginning and end of each trail, thereby measuring the length and direction of the drift – and hence the plate scale. Using at least two or three bright trails helps to produce an average that

Fig. 6.6 *PEMPro* pattern recognition to identify direction movements. The mount is driven in right ascension and then declination for a few seconds to indicate movement directions

can be computed by the software. You should hopefully see a resulting image scale that agrees with the image scale already calculated by *Maxim DL* during a plate solve (assuming that you did this as instructed in Chap. 5). Confidence in the process can be high when these two figures are seen to agree!

One stage of the calibration process identifies the direction of the telescope's drive for specific commands (see Fig. 6.6). Aim to have your CCD camera orientated so that star trails are along the X axis (RA axis). Clearly, there should be no vertical drift during this short period.

Using the Simulated Handbox

During the early development of *PEMPro*, Ray was asked about using the program with a newly acquired RCX400 scope, having previously used it to great effect with an LX200 Classic. Ray kindly made some additional modifications and added some new features. Consequently *PEMPro* also caters to the LX400 and provides the simulated handbox display that can be used to ensure that PEC is running and synchronized correctly.

Virtually all handbox commands are duplicated on this display. The program can also display the Hour Angle – a good indication of how far west the scope has driven. Most importantly, this display can be used to both switch on and synchronize the PEC table with the computer – thereby ensuring accurate PE measurements. This is most useful in remote networking mode because of the obvious convenience. When you have activated the handbox display you can switch PEC on (it may already be on) Toggle *High Accuracy Sync* (35 s) and click *Sync PE*. This process then accurately synchronizes *PEMPRO's* counter (as displayed on the screen) with the telescope's *RA PEC* segment counter.

Ray explained that to keep track of the current PEC Index *PEMPro* uses a starting timestamp that it saves when you synced the PEC index. To calculate an index in the future *PEMPro* subtracts the current time from the starting timestamp and multiplies that by 2.4 (seconds/PEC cell for an LC200GPS/LX400), and then adds the index saved at the starting timestamp. Finally, if the index is above 600, 600 is subtracted continuously until it is within the range 0–599. Whenever you move the mount (whether due to a slew or a nudge), *PEMPro* loses track of the actual PEC index of the mount, so the initial index is lost. Consequently, whenever you move the telescope to a new position, you must resync *PEMPro's* PEC index.

Handbox Settings

If you already have your LX200GPS or LX400 powered up, hold down the *mode* key for a couple of seconds and release. You can then scroll through various display settings and locate the *RA PEC* segment position. (There is also a *DEC PEC* facility, though this is not normally used unless you have a relatively poor polar alignment. For optimum operations, polar alignment must be improved.) If your PEC table is not active, the PEC segment will not be shown in this mode. You can activate it (once the handbox mode key has been held down) by selecting *setup/telescope/RA PEC on/off* and switching RA PEC on. When your PEC table is switched on, the current segment position can be displayed on the handset.

As mentioned, there are three possible settings for your PEC table:

Default: This is the way that your telescope is supplied (out of the box), with the PEC normally on. The PEC table has been prepared at the factory and contains a basic list of correctional data for the worm drive. This PEC has clearly not been prepared using your telescope with its fitted accessories and current balancing hardware, so the table's current data will differ to a greater or lesser extent from a newly measured raw PEC. At this stage, for your first image check with PEC activated, you should ensure that the PEC is on and therefore using the default table.

Zero: The next PEC content option is the zero setting, obtained by zeroing the PEC table. You will use this shortly when you are ready to measure the absolute PE of the scope. An important fact here is that in order to set PEC to zero, it must be switched on!

Optimized: The third PEC setting is the optimized one, prepared by measuring the actual periodic error (PE), computing a correction curve, and uploading it to the telescope to become the new PEC table. Until you have done this, continue to use the factory default. This chapter takes you through the process of measuring the raw PE and uploading a new correction curve.

Collecting Raw Data: A Caution

If you have completed the previous stages and found a suitable star for monitoring, you can shortly start collecting PE data. Ensure that you have a suitable measurement star just west of due south and something above zero declination – either where you ran the calibration during the previous procedure or manually when you are ready.

You are going to measure the periodic error of your scope, so the scope and its PEC table must be set in the correct mode. You do *not* want to measure the PE of the factory data, so you must deliberately zero the current PEC table in order to measure the raw PE of the scope. For the LX400 and LX200GPS, the current PEC must be set to ON and zeroed. Use either the real or the simulated handbox display referred to previously in this chapter for switching PEC on.

PEMPro permits you to download the current PEC directly from the mount. To do so, select the telescope tab (*LX200GPS* or *RCX400*) from the tab options. Select *LX200GPS from mount*, or for the LX400 RCX400, "*from mount*"; this downloads the current PEC (probably the factory PEC that was already present). You do not need the actual data, so click on *Clear PEC* to zero it (be aware that you are zeroing the PEC table now copied into *PEMPro*, not that on the mount). It is essential to zero the telescope's PEC table before any measurement is started, or you will merely measure the telescope's current factory PEC table! Select *To mount* and upload the zeroed PEC to the mount's PEC table. You now have a known zeroed PEC. For this purpose (producing a zeroed PEC) you do not have to download the current PEC; you can sometimes load an old one and simply zero it ready for use. This ensures that PEC is both on (activated) and zeroed.

Having zeroed the PEC, you can start the measurement. Click on *start*. *PEMpro* takes a full-frame image using your selected exposure and then attempts to find a suitable star for measurement near the center of the image. Assuming such a star is located, you should see a set of readings appear in the table and a graph displayed in real time, showing the relative movement of the error (compared with the first reading). Ensure that the maximum pixel level is not too low or too high. If too high, it might saturate, and then the algorithm that measures its centroid cannot derive an accurate position. Always use this first run as a test run just to check that everything is OK.

You can terminate this test run if you wish to change any parameter. It is a good idea to do one or two short runs to check that all is well and then start the main run. With both the LX200GPS and the LX400, you can start at any arbitrary time, because you have already synchronized the computer's reference start with the actual PEC table. Do remember to resync *PEMPro* if you move the scope. It is also worth resyncing *PEMPro* if much time has elapsed since the original time synchronizing.

During the long run – and remember that you are aiming for three complete cycles of the worm (about 3 times 24 min) – the scope will drive about 20° westward. You should therefore ensure that during this time, the aperture of the scope has free access to the sky and suffers minimal vibration. As each track reaches the right-hand side of the graphical area, the color changes. (It probably would have been better to have the color change at the end of each completed 24-min cycle.)

There are four box options used to change the display. You might want to have the display showing X and Y in arcseconds, so this box is ticked. When the first and subsequent cycles are complete – see Fig. 6.7 – the track will inevitably have

Fig. 6.7 *PEMPro* displays three full cycles of periodic error from an LX400 following a complete 72-plus-minute measurement

drifted either upward or downward, depending on the offset of the scope's polar axis from the NCP (the true north celestial pole). In another mode – the polar alignment measure – this offset can be identified and used (at another time) to reposition the polar axis in real time to improve tracking. For now, note that the *Adjust for Xdrift* and *Adjust for Ydrift* boxes can be ticked to change the display to counteract the drift by realigning the tracks. This display does not affect the data itself in any way, merely the display. It is where you can quickly see whether your mount is too far offset from the pole.

A glance at Fig. 6.7 tells you a lot about the telescope. Despite the author's brief efforts at polar drift alignment a year or two before these measurements, there remained some misalignment, shown by the slow drift of successive curves. The basic 8-min worm cycle can be seen, though with considerable modifications due to the other components of the gearing system. This result shows the value as well as the need to consider the three 8-min cycles as merely comprising one complete cycle. The overall peak-to-peak error of about 18 arcsec can be estimated visually.

Analyzing Your PE Data

This is where the fun starts! *PEMPro* can now produce automatic correction curves using this data. Select *Create PEC curve*. The accumulated data is analyzed using advanced Fast Fourier Transform analysis, and the resulting numbers are displayed

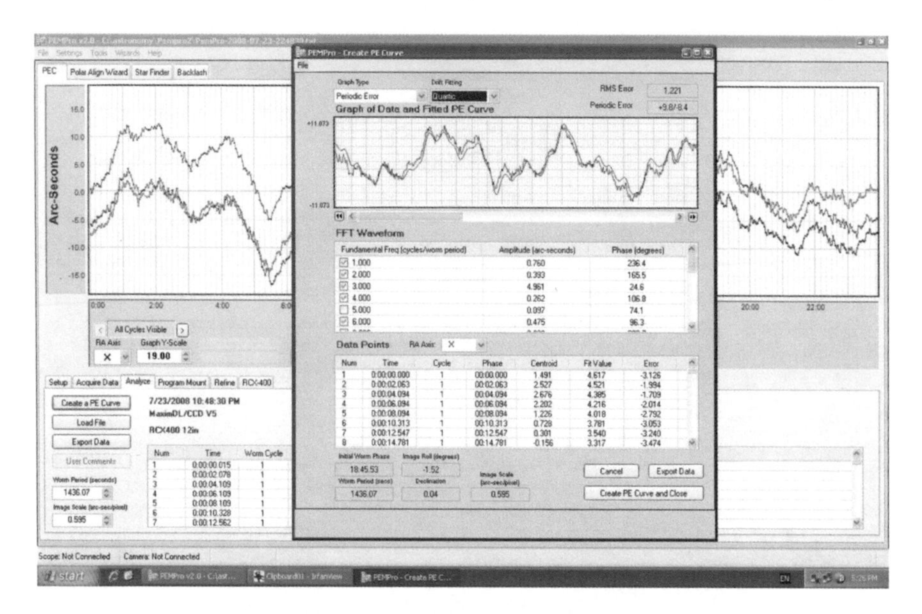

Fig. 6.8 *PEMPro* analysis of raw periodic error. The best fit curve can then be selected

(see Fig. 6.8) in a separate window. Hopefully your run will have been uneventful (no clouds interrupting the measurements and no cosmic ray hits), and the drift will have remained on the graph. Analysis can take several seconds because it is CPU intensive. The final results of the measurement are displayed at the top right, in the form of an RMS error and the Periodic Error, peak-to-peak. The former is a measure of how good a fit the computed curve is to the drift-fitting equation used for the computation. The resulting Periodic Error curve is your *first* measure of the raw tracking capability of your unmodified mount. The program offers a huge number of analytical tools so that if you are so inclined, you can really get to understand the intricacies of your mount's gearing.

The results can be displayed in various ways to help this analysis. You can simply let *PEMPro* do the job of producing a correction curve. If your long run included a few seconds or so of bad data, caused, for example, by a small cloud crossing through the star field and temporarily stopping monitoring, this event will be noticed during visual inspection. The program permits the editing of this data.

This is not the end of the measurement, however. *PEMPro* can repeat the entire calculation exercise using higher order algorithms, as indicated in the *drift-fitting* option. Higher order algorithms are likely to produce curves having a better fit (as indicated by the r.m.s. error reducing during successive calculation runs), though if you have collected insufficient data to validate these higher orders then the result is unlikely to improve. Because you have obtained effectively nine complete worm cycles (each 24-min complete cycle comprised 3 actual 8-min cycles) and you ran for 3 times 24 min (totaling over 72 min), it is perfectly permissible to go

Table 6.1 Results obtained using *PEMPro* on LX400

Linear fit	r.m.s. 4.705	+9.2/–9.2
Quadratic fit	r.m.s. 2.674	+8.7/–9.8
Cubic fit	r.m.s. 2.634	+8.8/–9.7
Quartic fit	r.m.s. 1.221	+9.8/–8.4
5th order fit	r.m.s. 1.042	+10.0/–8.6

above quadratic and use the higher orders, as shown in Fig. 6.8. Write down the results obtained using each order up to fifth order, so that the sequence of changing r.m.s. and PE measures can be appreciated. (See Table 6.1 above.)

Those interested in the complete analysis of the results can display the curves in terms of a Frequency Spectrum. *PEMPro's* help section provides an extensive analysis option for this, and the user can, if he or she so wishes, modify the result accordingly. When you have decided on the appropriate high-order algorithm (i.e., the quartic order result in this instance), select the option Create PE curve and Close at bottom right of Fig. 6.8. The result is the optimized correction curve ready for uploading to your telescope but on the *program mount* tab. However, both the LX400 and LX200GPS types of telescopes, and several others, allow direct uploading of the PE correction curve to the telescope's computer. You can therefore bypass the Program Mount option –which is essentially designed for the slow playback of data into the telescope – and select the last option, the telescope (*RCX400* in this case) tab itself.

Uploading the New PE Correction Curve

From the previous stage where you produced the curve, select the telescope (RCX400 or LX200GPS) tab and click on *From Created PE Curve*. This transfers the curve to the display. *PEMPro's* PE should be resynchronized to counteract any drift, and you can then use the option to send curves *to mount*, or *to a file*. Other options include retrieve the mount's PEC via *from mount*, or *from* a *file* already prepared. There are numerous options here, and all are described in detail in *PEMPro's* help section.

A cautionary warning is given for you to check the movement direction (settings) to ensure that it is set to *Forward*; otherwise the curve could be inverted without you realizing this. If you have completed the *wizard mount calibration*, all settings should be valid. Upload the correction curve by clicking *to mount*. There are other ways of uploading the curve, but these are best reviewed from the help facility.

Check the Result

It is now essential that a further several minutes of PE are collected from the telescope on a suitable measurement (guide) star in order to confirm that you really have improved the scope's periodic error. For a quick check, reposition the scope on a suitable star near the meridian as before, resynchronize the PEC clock (because you have just moved the scope nearer the meridian), and start collecting some new data. It will quickly become apparent if you have got it right or wrong! If the new periodic error curve swings quickly into bad territory, showing a large increase in both variability and error, then the curve might be inverted or some other error is having an effect. The most likely cause of a bad result during this quick check is that you did not perform the image scale and direction calibrations. It is always tempting to simply enter the data from a previous image solve, as mentioned earlier. However, doing that could leave a direction setting wrong, and that would dramatically affect the result. If you do get this bad result, go back to the previous correction curve and select *invert*. You can then upload this modified curve and check the new PE again – after resyncing *PEMPro* with the scope. If all the stages have been done carefully, then collect more PE data and the trace should be greatly improved.

Refining Your Curve

Results obtained by this author using *PEMPro* in July 2008 changed the original raw PE of 18.2 arcsec to 4.3 arcsec – a satisfactory result! *PEMPro's* help section explains in great detail exactly how to remeasure and produce a second correction curve, so for the purposes of this chapter, it was informative to do a refine again a few nights later. This requires a remeasure of the first level of correction for repeat three cycles (24 min times 3) – in other words, doing an action replay but using the corrected PEC. This time, the resulting curve is subjected to the *refine* option. By combining this curve with the original one, *PEMPro* can produce a second curve that can be directly uploaded to the mount.

Start the refinement process by setting up *PEMPro* as if for the first measurement. Find a suitable measurement star just on the west side of the meridian and close to the celestial equator (declination near 0°). Use the *focus* option to adjust the telescope so that the star is near the center of the screen and not saturated – exactly as you did first time around. Load the handbox display, activate the PE high-accuracy box, and then (as before) select sync PEC. When this is synched, the telescope's PE table is synchronized with *PEMPro* and you can start the measure. Figure 6.9 shows the result of measuring these three new cycles a day or so after the first level PEC correction was prepared. You can clearly see the significant reduction in periodic error since the correction curve was uploaded. The results of this refine process are seen in Table 6.2.

Selecting again (in this case) the quartic fit curve, you can produce a PE correction curve and transfer it to the *refine* option as curve 1. Load curve 2 by finding the previous correction curve produced during your first measurement session, as shown by the dates on Fig. 6.10. The curve with the largest variation is the original correction curve (curve 2); curve 1 shows the latest (small) correction calculations. The "add curves 1 and 2" feature (bottom left) produces the intermediate curve. You can save this as the *created curve (green)* and set the *program mount* option to

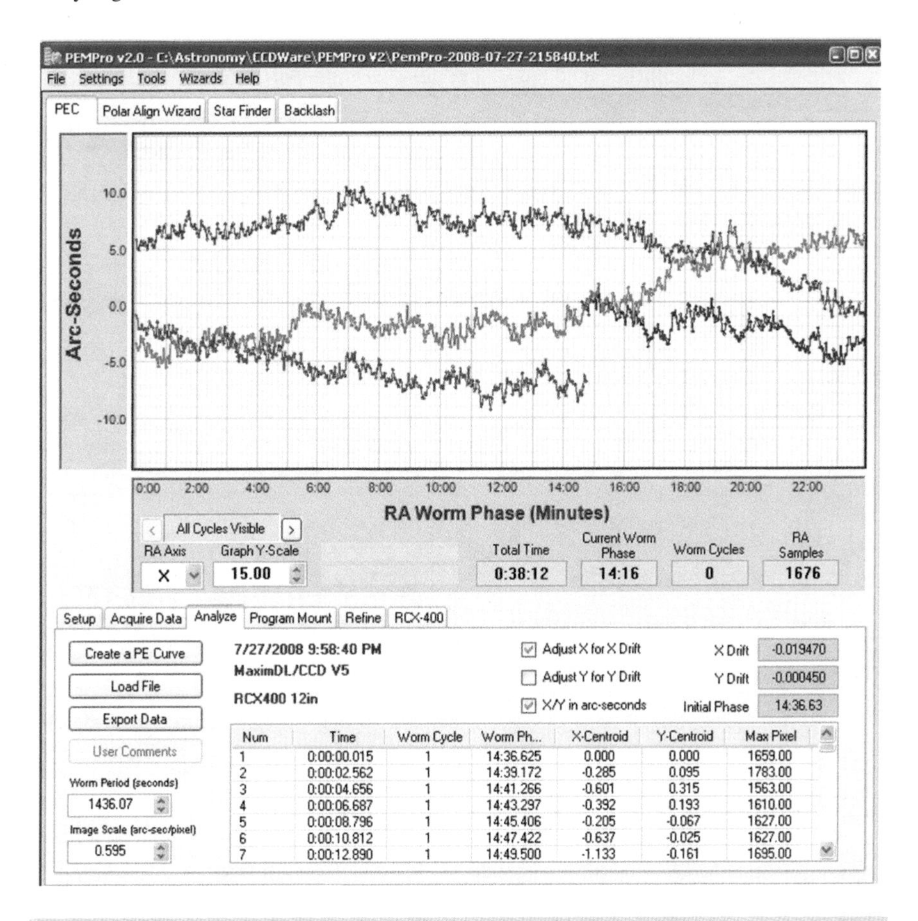

Fig. 6.9 *PEMPro* measures the new, corrected PE. Note change of vertical scale for display purposes. The variations here are much smaller than the uncorrected PE

Table 6.2 The new corrected PE

Linear fit	r.m.s. 4.843	+1.9/−1.9
Quadratic fit	r.m.s. 2.837	+2.5/−2.0
Cubic fit	r.m.s. 2.046	+1.8/−1.7
Quartic fit	r.m.s. 1.297	+1.2/−1.2
5th order	r.m.s. 0.853	+1.3/−1.0

upload it. This curve is the optimum correction. Before uploading it, do another resync of the *PEMPro* table with the telescope's PEC table in order to ensure synchronization. Then upload.

Once more you should do a single-cycle PE measure to check that all is well. Figure 6.11 shows a final result. With a final PE indicating about ±1 arcsec, this can

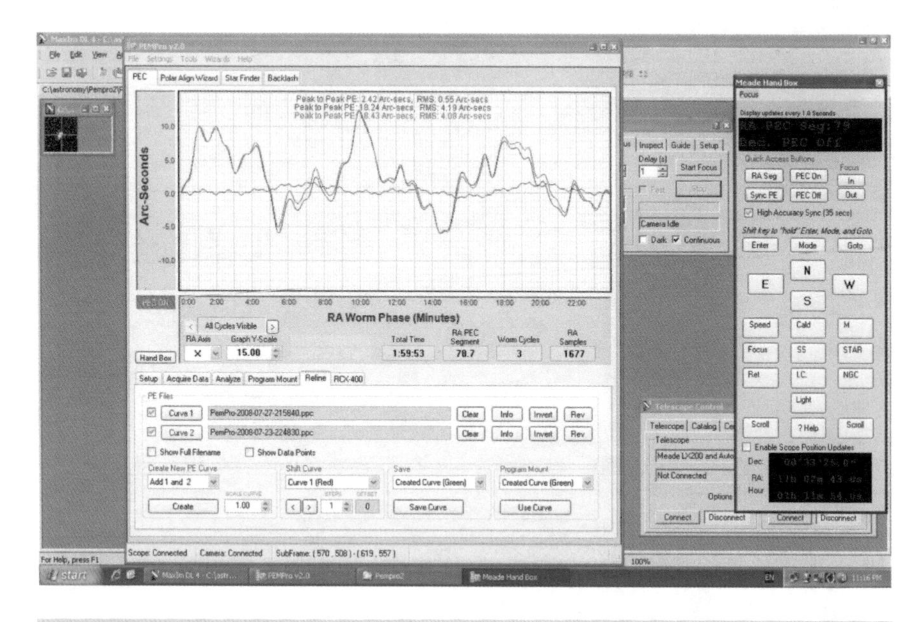

Fig. 6.10 *PEMPro* refine curve – combining two correction curves to produce an optimum correction

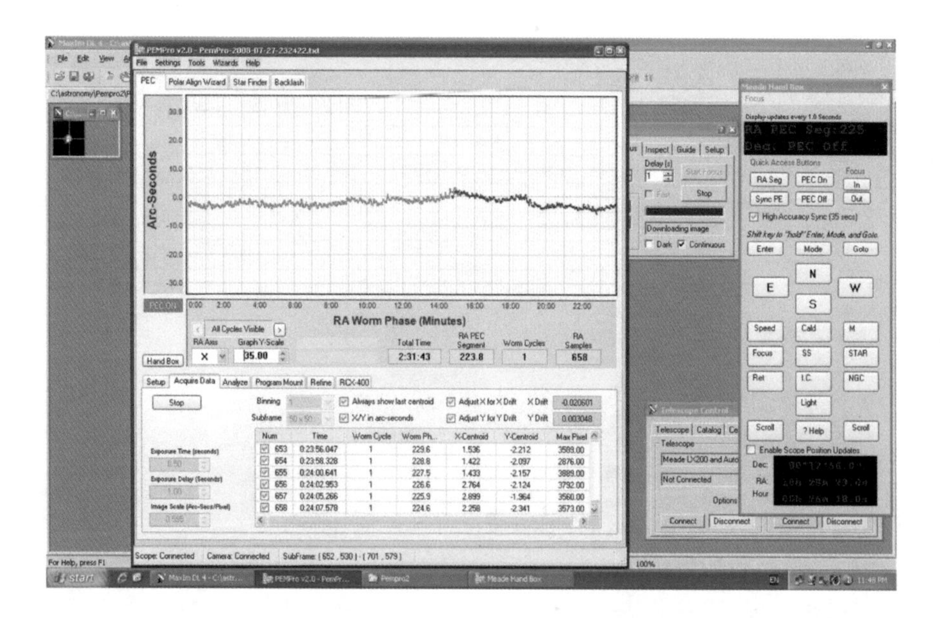

Fig. 6.11 *PEMPro's* final PE measure following both correction runs

surely not be improved! This result is highly satisfactory, and by maintaining a record of the PE measurements and measuring again every few months, you can monitor any degradation in the overall performance of the periodic error. The result compares with the magnitude of atmospheric scintillation. When your own measurements indicate that your final PE is acceptably low in comparison with the raw measurements, you have improved the tracking capabilities of your scope far beyond its out-of-the-box state.

What a Good PEC Means

When the periodic error of your telescope has been reduced to a few seconds of arc or less, it should be capable of tracking really well. This means that if you locate a target in the CCD frame and leave the scope tracking without any further intervention – such as autoguiding – the target should remain reasonably centered within the frame for some minutes. There is likely to be a slow drift of the target either down, up, east, or west within the frame during a period of several minutes if the polar axis is not well aligned. This is why it is important to have a reasonably good polar alignment before running *PEMPro*. As you have seen, *PEMPro* can compensate for a moderate amount of polar misalignment.

Adjusting the Sidereal Drive

Our atmosphere exerts a profound effect on the images produced by telescopes. Because of refraction, the elevation of a star (or other astronomical object) is seemingly increased so that we see it slightly higher in the sky than its true position. We therefore see objects for rather longer than is theoretically possible, because they are visible before they theoretically rise above the horizon. This is a well-known effect, but it has a consequence that is not always appreciated. When looking through the telescope at an object, the object appears to move slightly more slowly that the scope's built-in sidereal drive. Looking at it the other way, the scope is driving at a slightly higher speed than the star's movement, so there will be a residual drift during almost all monitoring.

The LX series scopes offer an adjustable drive rate: select *setup/telescope/tracking rate/custom*. Before every observing session the author sets the custom rate to –001. The effect of this reduction can be seen in Figs. 6.12 and 6.13.

Notice that in each figure (6.12 and 6.13) the *Adjust X for drift* setting is off so that the true RA drift will be shown. The custom RA tracking rate has to be set each time that the scope is powered on because it is not retained. After power on you should activate the dew heater (at least during autumn, winter, and early spring) and then reduce the tracking rate as well.

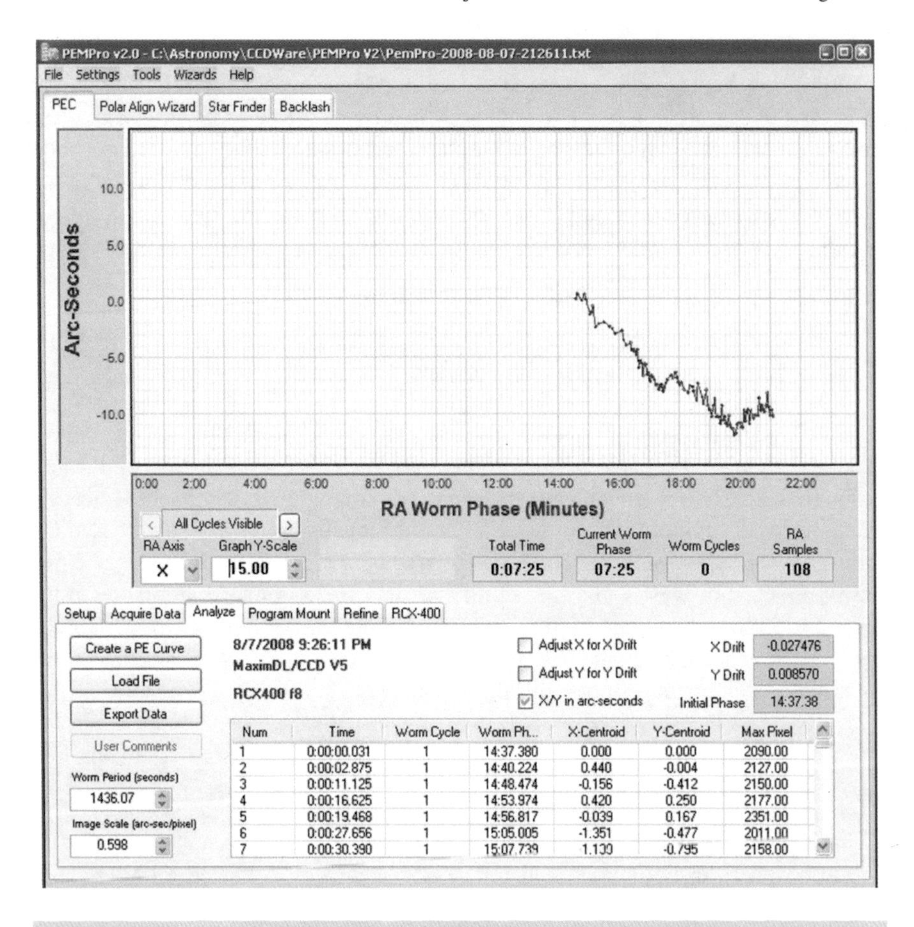

Fig. 6.12 Monitoring PE before adjusting the RA drive rate to demonstrate drift

Polar Alignment: The Importance of Being Accurate

This topic has its own dedicated section in Chap. 4, but a brief summary here is appropriate because *PEMPro* can measure it. If the scope is not already approximately polar aligned, any stars, or other targets, are likely to drift north or south across the field of view. An approximate alignment can be achieved fairly rapidly, as explained in the section. With poor polar alignment, a full 24-min PE cycle measurement would see drift that would probably take the star out of the field of view. Because you should aim to measure three cycles (3 times 24 min) in order to further improve accuracy, good polar alignment is essential. *PEMPro* will automatically compensate for a nominal amount of polar misalignment, but now includes a polar alignment feature to both measure and help correct any offset.

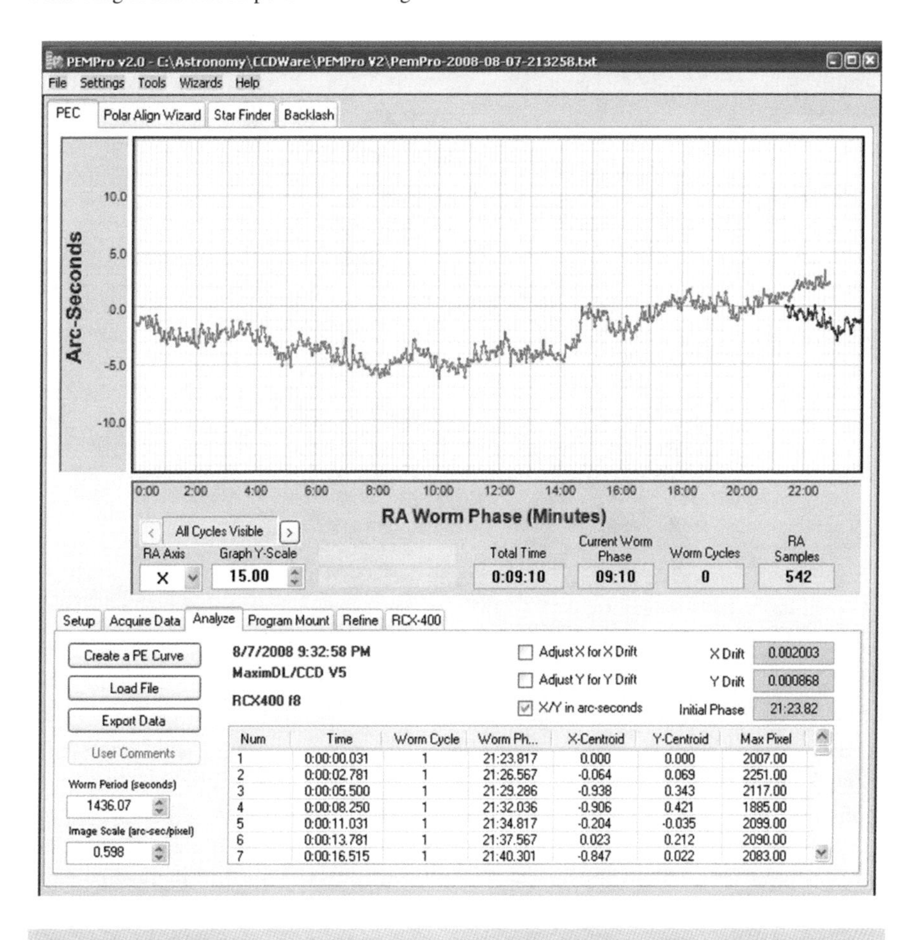

Fig. 6.13 Monitoring PE after RA drive rate reduction. Note the considerable reduction of drift

Software Polar Alignment

The *PEMPro* Polar Alignment feature provides some direct measurements to assist polar alignment. Setting up the software parameters within *PEMPro* was described prior to the measurement of your telescope's periodic error; these settings remain valid if your camera is still set up the same way. If you have not set up the software, you will need to follow the procedures described; these involve defining your scope's parameters and those of your camera. Use the Polar Align Wizard; this offers a step-by-step approach to configuration. The wizard slews the telescope to a suitable position according to your preferences for the start of the procedure. The image scale is then measured, as described in the earlier section on periodic error. Trail beginnings and ends are marked to enable the calculation of camera orientation to be made.

Fig. 6.14 Ray Gralak, creator of *PEMPro*

By selecting the Polar Align Wizard you can repeat the process of selecting a suitable measurement star for monitoring purposes. As before, *PEMPro's* help facility is exceptionally comprehensive, taking you from initial configuration of the software, telescope, and camera through to selecting a suitable star. Find a suitable star first and then enter the wizard. The program makes the assumption that the mount is nominally polar aligned already. If you followed earlier suggestions about locating, constructing, and setting up the wedge, then this rough alignment will have been done. If the mount is within 5–10° of the north celestial pole, then you are ready. Our earlier suggestions about preliminary alignment should have left it well within this level.

Having found a suitable star of moderately bright magnitude you should synchronize the telescope to its position. Ensure that details such as site location, etc., as required on the site information page, are entered accurately. Unlike when monitoring periodic error over long timescales, these measurements will generally last a matter of a few minutes at a time. The camera's image scale needs to be known, and its measurement is calculated during an early stage. A series of images are taken to positively identify the cardinal directions (north, south, east, and west). The first measurement made is that of azimuth – also the easiest. When ready, the measurement of drift can commence.

The process starts by showing the developing trend that should stabilize within a couple of minutes or so. When stability has been achieved, a suggested azimuth movement will be displayed. Note which direction and then make a movement – not necessarily the whole amount! Allow a further drift measurement to occur;

hopefully the magnitude of the required movement will have been reduced, showing that all is well. If the trend worsens, try changing the azimuth in the opposite direction, and then remeasure the new trend. This will almost certainly produce an improvement. By continuing the process of measurement and adjustment your polar alignment should be considerably improved.

It is worth spending the time to make use of the periodic error correction feature because of the significant improvement demonstrated here. You should measure your current PE on a regular basis (without zeroing the PEC table) just to ensure that your scope continues to track well. Ray Gralak, creator of *PEMPro* (see Fig. 6.14), was very helpful for specific advice during the writing of this book. His contribution is appreciated.

Further References

Andrew Johansen, Melbourne Australia: http://members.optusnet.com.au/johansea/
Look for the *MyScope* file, download, unzip, and run (say "no" to the scope searching question!). This document tells you everything that you could wish to know about the innards of the PEC data storage. Look at the beta file for later information.
Doc G's Information site about LX200GPS telescopes: http://www.mapug-astronomy.net/ragreiner/index.html

Chapter 7

Previous chapters have described how to optimize your telescope's balance, minimize backlash in both axes, obtain good polar alignment, and minimize the mount's natural periodic error. It really is important to optimize all of these parameters before serious attention is given to autoguiding, for reasons that will become apparent in this chapter. By now, even without autoguiding you should already be able to take a series of pictures with your CCD camera, exposing each for at least 1 min, preferably rather more, and expect the majority to be decent images with round stars.

Because care has been taken with PEC, it should be possible to extend exposures for possibly 2 min – or even longer – before drift from any cause spoils the round stars, making them elongate. A glance at images of unguided exposures – see Figs. 6.2 and 6.3 in Chap. 6 – shows the basic results that should be typical for both the LX200GPS and LX400 series of telescopes. Autoguiding should now enable you to use much longer exposures, lasting 10 min and more. In all cases, you are striving to obtain individually longer exposure images of a quality matching the shorter-exposure unguided images.

Principles of Autoguiding

The main principle of autoguiding is straightforward. The usual method involves placing a second telescope on top of the main telescope (see Fig. 3.1 in Chap. 3). The guidescope must be fitted rigidly to the main telescope. The guidescope is fitted with a small guide camera and takes a succession of short-exposure images of a suitably selected guide star. This star is likely to be close to the actual target of the main telescope because both scopes are normally pointing in the same direction. During a long exposure on the main camera, the autoguiding software detects small changes in the exact position of the image of the guide star (and therefore the target)

L. Harris, *So You Want a Meade LX Telescope!*, Patrick Moore's Practical Astronomy Series, 127
DOI 10.1007/978-1-4419-1775-1_7, © Springer Science+Business Media, LLC 2010

and commands the main scope to make small compensating corrections. These corrections are made after the analysis of *every* guide star image, so they may average about one correction every 2 s or so. Assumptions yet to be explained are made with this process, and their validity is a factor in its efficiency – or lack thereof. We shall see how many such autoguiding combinations can be further optimized to improve the results of autoguiding.

Why would there be small changes in the position of the guide star between successive exposures? Despite all optimization efforts, no mount is perfect. Several factors can influence the position of the guide star image:

1. Residual periodic error. There will still remain a level of periodic error, however small.
2. Polar alignment errors. Unless perfectly polar-aligned, there may still be a small residual drift in declination due to this misalignment.
3. Refraction changes. Unless you are viewing at the zenith, there will be refraction changes due to the changing altitude of the target object. Lower-altitude objects are refracted more through the thicker atmosphere than are higher-altitude objects, so this refraction level changes all the time.
4. Atmospheric scintillation. If you were to take subsecond images of a star, you would see the star moving about over a small radius, quite randomly, as air cells are affected by local atmospheric thermals. This form of autoguiding is not normally able to correct for such very short timescale changes. In a later chapter, it will be explained how you *can* deal with this at additional cost – if you so wish. Meanwhile, all instantaneous images of the guide star may well provide a slightly erroneous measurement because of this variation.

All these changes affect the position of the guide star image, so the software will compute the instantaneous difference and send a suitable correction to the mount. So, in theory, the following sequence occurs:

1. A short exposure (perhaps between 1 and 5 s) is taken of the guide star by the guide camera.
2. Software calculates the position of the centroid (the brightest pixel within the guide star).
3. The guide camera takes another image.
4. The software calculates the new centroid position.
5. The software calculates the command sequence required to move the mount to realign the new centroid with the original (this calculation is made using the guide camera calibration).
6. The sequence is repeated, in effect always moving the mount so as to realign the latest centroid with the first guide star centroid.

Because no mount is perfect, there will always be a need to correct for the various imperfections previously described. There are further, significant complications that make the whole process more than merely a textbook exercise.

1. All commands sent by the autoguiding software to the mount rely on the original guide and mount calibrations, but these are themselves not 100% accurate or

even consistent. Two successive calibrations are unlikely to produce exactly the same results because of reasons explained earlier. Consequently, the actual commands sent by the software to the mount are unlikely to be 100% correct, and the mount itself may not react consistently to them.

2. There is a built-in delay between the measurement of the most recent position and the delivery of the command calculated to be required to move the mount to the correct position. When the mount actually completes this correction, the required correction is unlikely to be the true amount any more.

Backlash

The final command issued to the mount may be a combination of declination and right ascension movements. Polar misalignment may cause a slow drift in declination, and periodic error may produce apparent changes in right ascension. Both axes will exhibit backlash to a greater or lesser extent. Backlash is the condition in which a change of direction commanded to the mount causes the motors to reverse direction, but the gears have to remesh due to manufacturing imperfections. This backlash should be largely minimized during the earlier session of backlash reduction described in a previous chapter, the capability for which is provided by the Meade telescopes.

Other factors may produce added complications to the autoguiding: friction (sticky axes), stiction (static friction), which is a perverse event in which a command to move in one direction causes an initial lurch in the opposite direction, and the astronomer's worst nightmare – poorly routed cables! If the axis is unable to move after the command (due to, for instance, a cable getting stuck), the energy remains available for release. When the energy has built to a critical level, the mount will suddenly move. By this time, repeated – and therefore unwanted – corrections will have been measured and sent to the motors, which will probably overcompensate for the actual movement required. This overcorrection may continue if the autoguiding settings have not been optimized. It is therefore important to optimize your autoguiding, so that it needs the minimum number of guide corrections.

Autoguiding Error Magnitudes

It is worth giving some thought to the likely magnitudes (sizes) of the errors that could crop up from some of the potential causes of guide star drift.

1. Polar misalignment. If the mount has been well polar aligned, the drift amount should be quite small. During a 10-min exposure, it should only amount to a few arc seconds. Significantly, it should only be in one direction.
2. PEC. This should also have been minimized during the PEC measure and correct process, so the maximum PE error during a few guide cycles should be incredibly small.

3. Seeing. Thermals are widespread, particularly during the early evening, and these can play havoc with autoguiding. There must be very few astronomers who can rely on calm atmospheric conditions every night. Excluding other factors, seeing will cause the guide star's position to move randomly (oscillate) around its true position, in a circle of possibly 1–3 arc s.

Clearly, during a typical guide cycle (say 2 s), the drift and PE errors are tiny in relation to seeing variability. If your autoguide settings are not carefully selected, you can be forcing your autoguider to chase the seeing because that is the main cause of any movement between successive autoguider star images. This would result in images being completely spoiled due to smeared stars. In Jim McMillan's seminal article on autoguiding (see References at the end of this chapter), he shows that seeing variability can be 20 times the magnitude of PE and drift variability.

Getting Good Guide Stars: Choose Your Guide 'Scope Carefully

MaxIm for autoguiding is one of a number of software programs that include an autoguiding function. To be able to start guiding and to guide efficiently, *Maxim* has to identify a suitable guide star. The success of this depends entirely on both the guide 'scope and the CCD guide camera.

The optimum guide 'scope (cost, weight, and performance) is likely to be much smaller than the main scope. Its focal length can be considerably shorter than that of the main 'scope, because modern cameras and software can guide on fractions of a pixel. If the guide telescope and guide camera combination produce an adequately resolved selection of stars, autoguiding should be feasible. Let us look at a typical example:

An LX400 operating at f6 gives an equivalent focal length of about 1,400 mm. The 80 mm *Skywatcher* guide telescope has a focal length of 550 mm. With this combination, you should normally be autoguiding to subpixel accuracy.

Flexure

The autoguider's main problem is usually flexure, the (possible) relative movement of the guide 'scope with respect to the main 'scope during an extended period of imaging. To minimize (if not totally eliminate) flexure, consider these suggestions when installing your guide scope:

1. Mount the guide scope on a solid, rigid rail system (see Fig. 3.1, Chap. 3).
2. Try to mount the guide 'scope close to the telescope's center of gravity. Mounting it too high on top of the OTA will amplify any flexure in your system.
3. If possible, mount the guide camera securely to the guide scope rail; try to avoid it dangling from the focuser tube.

As an example, the author's main scope is used at f6 (fitted with the Celestron f6.3 reducer) rather than its native f8 (2,400 mm focal length) because of the faster focal ratio and wider views. The 80 mm guide telescope shown in Fig. 1.5 (Chap. 1) is fitted with an SXV guide camera that gives an adequate guide star resolution for autoguiding purposes at its focal ratio. It is also within its capabilities to work with the main scope at its full focal length of about 2,400 mm, at which autoguiding becomes more challenging but by no means impossible.

Bright Guide Star

One of the main requirements for effective autoguiding is the availability of an adequately bright guide star. As long as it is bright enough – that is, has a sufficient signal-to-noise (SNR) ratio – *Maxim* can apply its centroiding algorithm and produce an accurate position for the brightest pixel.

What about hot pixels from the guide camera? Aye, there's the rub! Great care has to be taken with many guide cameras, especially noncooled ones, to accommodate hot or even warm pixels because these may be numerous and can completely swamp the genuine hottest pixels of your guide star. *MaxIm's* basic autoguide module (at least within the versions tested by this author) does not distinguish between a hot pixel (often a single pixel) and a genuine star (normally extended over two or more pixels); however, it does include a feature to auto subtract a dark frame and/or bias frame when using the standard autoguiding module, so if prepared correctly, there can be a considerable improvement in the detection of guide stars.

The actual operational process of achieving this is slightly more complex in the case of using the guider in the adaptive optics unit, as we describe in more detail in a later chapter. If at all possible, ensure that the guide star is bright enough to produce a valid autoguider image of the star with associated slightly darker pixels adjacent but preferably not saturated. The brightest pixels may be seen moving around within this image, vividly demonstrating the changing seeing conditions.

Guide Camera Exposure

The shortest exposures, those between about 0.5–1 s, will often produce only low magnitude guide stars (therefore having a low SNR) and might also produce too rapid a mount correction for any significant drift components. Longer exposures should reduce the amount of atmospheric scintillation variation per guide cycle (the exposure plus download time), thereby reducing the unwanted corrections.

The optimum exposure is likely to vary between sessions, and can only be increased to a certain extent. Seeing variations can vary widely, even when the exposure time is increased. To have a significant effect on the average guide star's SNR, guide exposures might have to be of the order of perhaps 10 s. Even then, changes due to PE drift could remain lower. Consequently, although the guide star's SNR will be improved, the law of diminishing returns applies.

Gusts and Mirror Flop

At least two other sources of error may arise during an autoguiding session. Gusts of wind are likely to have a dramatic effect on telescopes that are not in some form of observatory and therefore not shielded from the wind. Wind buffeting is clearly a very short term error and should therefore *not* be corrected for.

The design of the LX200GPS series 'scopes is such that small movements of the mirror can occur, particularly when the meridian is crossed. This effect is referred to as mirror flop. Recent versions of the LX200GPS have included a locking device, as mentioned in Chap. 2, that helps minimize this movement. The LX400 has a completely different mirror cell design that firmly holds the mirror in place, preventing any such movement. Any mirror flop in the LX200GPS series does, therefore, need to be corrected during guiding and will be experienced as a sudden permanent movement of the brightest pixel.

For guiding, you need to use a reliably bright (and therefore good SNR) guide star, but it must not be so bright that its center is completely saturated (therefore preventing the centroiding algorithm from working). You also need the ability to correct for any moderate, rapidly occurring guide errors that might be caused by physical anomalies.

Correction Commands

If you have previously optimized your mount and have established that you can get good results on most nominal exposures – even without autoguiding – then that is excellent. At first sight, you might anticipate that guiding software should send a full correction command (100%) to the mount. In fact, doing this would be likely to result in oscillations. The result could be smeared images due to the reasons previously discussed – namely that the *true* correction amount is very unlikely to be 100% accurate. Common practice when autoguiding is therefore to reduce the *aggressiveness* setting (the proportion of the 100% to be applied) to between 50% and 80%. We shall see later how this works.

In Jim McMillan's paper on autoguiding (referred to earlier), he introduces his concept of mount guiding S/N ratio, which he defines as:

Mount Guide Signal = mount moves (arc second) made that *need* to be made.
Mount Guide Noise = mount moves (arc second) made that should *not* be made.

Clearly, in order to achieve the best possible guiding results, you have to maximize this mount guiding S/N ratio. For a given telescope setup, the guide exposure length significantly affects the numerator. You also have to minimize the number of bad corrections made, and the effect of making these bad corrections.

Bad corrections result from the following already known sources:

1. Physical problems, such as loose cables, mechanical problems between components, mount mechanics, and wind gusts that could cause tracking problems. Fix these where possible!

2. Poor polar alignment; this will result in more guide corrections.
3. An uncorrected periodic error (see Chap. 6).

Software guiding parameters clearly need to be set carefully in order to minimize the number of those guide corrections that are largely resulting from noise, and to try to minimize their effects. Follow these rules:

1. Ensure that the guide star's SNR is adequate to enable the centroiding calculation to be consistently accurate. Remember that the algorithm cannot differentiate between a central saturated portion, where all the pixels are at maximum, and from a grossly under-exposed star. Identify a guiding exposure that is long enough to maximize SNR yet short enough to send valid corrections often enough.
2. Recognize the need to optimize guide parameter settings in order to minimize the number of bad corrections.

MaxIm DL Guide Settings

Maxim DL has been used by many for autoguiding for some years, and other guiding software is likely to include comparable guide setup parameters. These parameters are explained now.

1. *Aggressiveness.* The percentage of the calculated movement that will be applied. Setting 10 = 100% (full correction), setting 0 = 0% (no correction).
2. *Maximum move.* The largest detected move that should be corrected.
3. *Minimum move.* The smallest detected move that should be corrected.

The parameters *maximum* and *minimum move* are in seconds, and are specified to allow us to limit the maximum move correction that can be sent to the mount, and also the minimum. You can now see the importance of being able to set these at levels that minimize unwanted corrections from short-term events. The *minimum move* setting offers the ability to limit corrections that might be due to minor centroid changes caused by PE errors.

As an example of a typical pulse correction, 0.100 represents a 100-ms pulse sent to the mount. The mount will be driving at sidereal guide speed, and such a pulse would produce a 1.5 arc s move, which is tiny. How likely is it that this calculation is exactly correct? For one guiding cycle, it is unlikely to be 100% correct for the reasons previously discussed. Logically, therefore, you do not want to send the full pulse correction to the mount. The validity of the software's calculated amount depends, as discussed, on the guide star's SNR, yet you do not want to be changing the settings according to the guide star's SNR every time you start a new exposure!

Consequently, a good starting setting for the *aggressiveness* number could be 50 (meaning 50% of full correction). This is a starting point. If you had absolutely no idea where to start when setting up your guiding software, there could be a temptation to assume that 100% must be correct. Our look at the potential sources of measurement error shows that this is very unlikely to be true. However, despite all the possible problems, the fact is that autoguiding does work – as regularly demonstrated by many experienced astronomers – and the parameters offered by *MaxIm DL* and other autoguiding software permit the optimization of these settings.

At all times, you need to avoid excessive corrections that could set the telescope oscillating – overcorrecting one way and then overcorrecting the other way. That is why some experimenting with settings is essential. It is not very time consuming, as will be seen.

The Importance of the Aggressiveness Setting

There are only two possibilities here: the setting is too low or too high. Let us suppose that 50% is actually too low. This results in the correction being insufficient to move the telescope by the correct amount. The subsequent guide star measurement is likely to show this and cause a second correction to be sent – also potentially too small. You should remember that the precise correction depends significantly on the guide star's SNR – unknown to you – and this may be low. Under those circumstances, and they are commonly experienced, the noise component of the measurement could be large, producing an unwanted large correction. In choosing a nominal (e.g., 50%) aggressiveness setting, you are reducing potentially wrong corrections by 50%, a significant error reduction. Any possible oscillations are immediately reduced in size by 50%. If the correction is relatively accurate, then the shortfall will be further made up during the next guider cycle. Any genuine shortfalls in guide correction movements should be made up within a few guider cycles.

The Maximum Move Setting

As previously explained, there are some genuine occasions when a large move would be valid (such as an instance of mirror flop), and some occasions where such a move would be invalid (such as a gust of wind). The probability of a genuine sudden large move being actually required seems low, so for best results this parameter should be set on the low side. After all, an instance of mirror flop is likely to be rare, and if it does occur, the change would be accommodated within a few guide cycles. If this does occur, the final image may well be spoiled by extended stars, but frankly this would probably happen regardless if the mirror is able to rock significantly. Gusts of wind will be ignored by a low setting, which will also help minimize the generation of oscillations.

The Minimum Move Setting

Remember that small changes in the position of the guide star centroid are likely to result from seeing variations rather than PE drift. If such small moves are ignored, the mount guiding S/N ratio is increased. Consequently, it is strongly recommended that you ignore all corrections that are smaller than the seeing variation.

This will obviously reduce the number of guiding corrections sent to the mount. Care must be taken, however, to avoid setting this parameter too high, in order to avoid ignoring corrections that really need to be made. The declination axis is the one most likely to suffer from some level of backlash, even after that has been minimized, so the setting needs careful experimentation to ensure that it is optimized.

DEC Compensation Setting

Movement in RA varies from being a maximum at declination zero (0°) to becoming zero at a declination of 90° (the North Pole) in the sky. The object of the *DEC Compensation* parameter is to maintain the same amount of RA movement relative to the sky, regardless of where the mount is pointing in declination. As we increase the declination of the target, this movement therefore reduces.

If *DEC Compensation* is *on*, each RA axis correction is increasingly likely to start chasing the seeing as the declination increases. The probable outcome would be oscillations in RA. The recommendation, therefore, is to calibrate the guider on a star at DEC = 0° and disable *DEC compensation*. In Jim McMillan's paper, as well as making this interesting and seemingly counterintuitive suggestion, Jim suggests trying an interesting experiment:

Calibrate your guider at dec = 0°, guide at dec = 0°, and also at dec = 60° with dec compensation first on and then off. You can then analyze the guiding logs to see whether turning on dec compensation helps or hinders the guiding results.

LX200GPS and LX400 Series PEC

Many mounts are known to have a raw PE characteristic in which the PE can be both fairly large (well over 10 arc s) and occur over a short period of time. It is therefore essential, as previously discussed, to ensure that PE is considerably reduced in order to ensure that all PE changes are both nominal (small) and occur slowly and evenly. Our look at optimizing the autoguiding parameters has (hopefully) shown that you can adjust the settings to minimize those errors that you can identify as being within known amounts.

A Mini Review

- On any given night there will be a certain level of seeing noise; you do not want to make guide corrections that are within the limitations of this noise. Consequently, there should be a number of guide cycles when no corrections are sent to the mount.
- Minimal oscillation: the settings should ensure that it only takes one or two corrections in the same direction before returning to a steady state.

- If you see oscillations, it means that there are corrections happening in opposite directions during a series of consecutive guide cycles. This is probably caused by chasing the seeing. Try increasing the *guide exposure* or try decreasing the *aggressiveness* setting, increasing the *minimum move* setting, or decreasing the *maximum move* setting.
- There should only be occasional corrections in declination, and these should be in just one direction, to compensate for any drift. A declination backlash problem will exacerbate oscillations. If there are declination oscillations, these are probably caused by chasing the seeing. Try increasing the *guide exposure*, decreasing the *aggressiveness* setting, increasing the *minimum move* setting, or decreasing the *maximum move* setting.
- If you find that a number of corrections are required in the same direction before the tracking error graph returns to a state where it does not need to make any corrections, it implies that the number of guide cycles where a correction can be made can be increased, so lower the *minimum move* parameter, or increase the *aggressiveness/maximum move* setting.

Guide Log Graphs

The software used for autoguiding (including *MaxIm DL*) can provide logs of guiding data following a guiding session. Graphs showing all RA and DEC corrections made during an exposure, using a known guide exposure, should reveal whether your guiding is good or not. If you define good as those producing almost no oscillation, almost no seeing chasing, no large physical anomalies, and finding that the majority of guide cycles did not require a guide correction at all, then that is good!

Sample Guide Settings

Guide settings are usually the result of your own telescope's physical setup and typical seeing conditions. They may or may not be appropriate for others since their situation is unlikely to be identical to yours. However, here are the details of the author's normal setup:

Mount: 30 cm LX400 f/8
Imaging camera: SXV-H9C
Image scale main camera: 0.65 arc s/pixel (highest resolution)
Guide scope: 80 mm tube, operating at focal length = 550 mm
Guiding camera: SXV guider
Seeing: rarely allows FWHM less than 2.5 arc s; usually around 3.0 arc s

MaxIm Guiding parameters:
  Aggressiveness: 5
  Maximum Move: 0.1 s
  Minimum Move: 0.040 s

DEC compensation: Off
DEC backlash compensation: 0.60 s (specific to my mount)
Guide exposure: Never less than 1.0 s.

With a short focal length guide scope, you will rarely need to increase the guide exposure above 1.5 s to achieve a sufficient guide star S/N ratio. However, two points arise. If the seeing is such that you need to increase the guide exposure above 1.0 s to achieve reasonable guiding S/N, it is probably not a very good time to image at high focal ratio (f8 or f10). As discussed, it is considered to be good practice to expose the guider for longer than this, say, 1–3 s.

Configuring the Autoguider

A previous chapter has a discussion on the setting up of *Maxim DL* with your computer and telescope. More advanced applications such as *ACP (Astronomer's Control Program)* require extra software, but for now, we shall look at the basic setup.

With the telescope running normally and the CCD camera system powered up, both USB cables can be plugged into the computer. *Maxlm* requires that both scope and cameras should be powered and their USB connectors plugged into the ports before *Maxlm* is started. You can then start *Maxlm DL*. When you are regularly going to use both 'scope and CCD with *Maxlm*, it is convenient to set up the toolbar with both 'scope and camera icons for easy start. You can then activate the 'scope module and then the camera connection module (or, of course, vice versa). The correct 'scope drivers must be selected, as described in the earlier chapter. If it is correctly set (LX200GPS for both LX200 and LX400 scopes), you can click connect. After a few seconds, connection to the telescope should be complete, although a pseudo error message may be displayed, claiming that an older version of the telescope driver has been detected. Before the camera *connect* button is selected, the camera must be configured. Ensure that the correct main camera chip is selected, then use the *autoguider* tab and ensure that the camera model is defined. Then click *connect*. The *Maxlm DL* manual explains about all of the settings. All being well, the *connect* tab should activate.

Focusing and Centering the Autoguider

The guide 'scope should produce a fairly wide field of view on the guide camera CCD chip (see Fig. 7.1). A good procedure for focusing and centering is to drive the scope to any bright star (use *Maxlm's Catalog* option and select a suitable bright star). The autoguide 'scope should include this bright star in its field of view – if you have previously centered this 'scope on the center of the main 'scope's field of view (see Fig. 7.2).

If you did not previously center the guide scope, it is an excellent time to do so now. This should ensure that wherever you drive the main scope, your guide camera provides a comparable view. Select *focus, guider* on the CCD module; a short

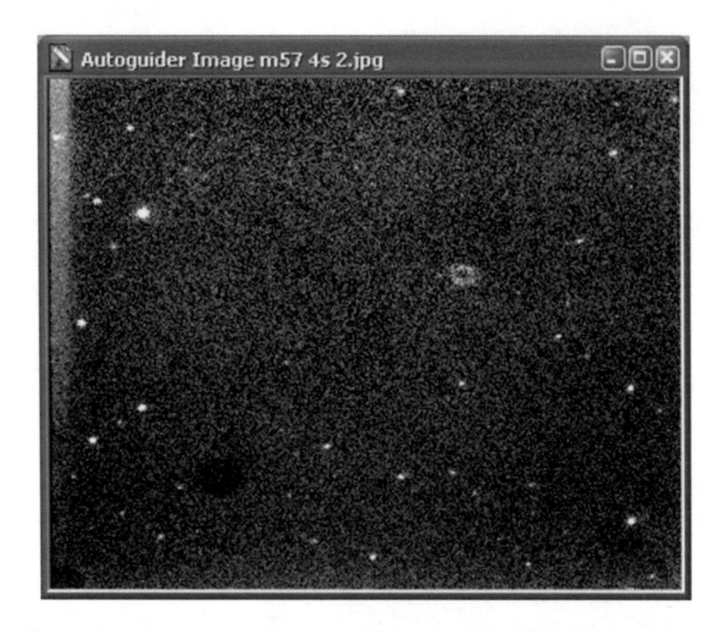

Fig. 7.1 Guide camera wide-field view of 4-s exposure showing Messier 57

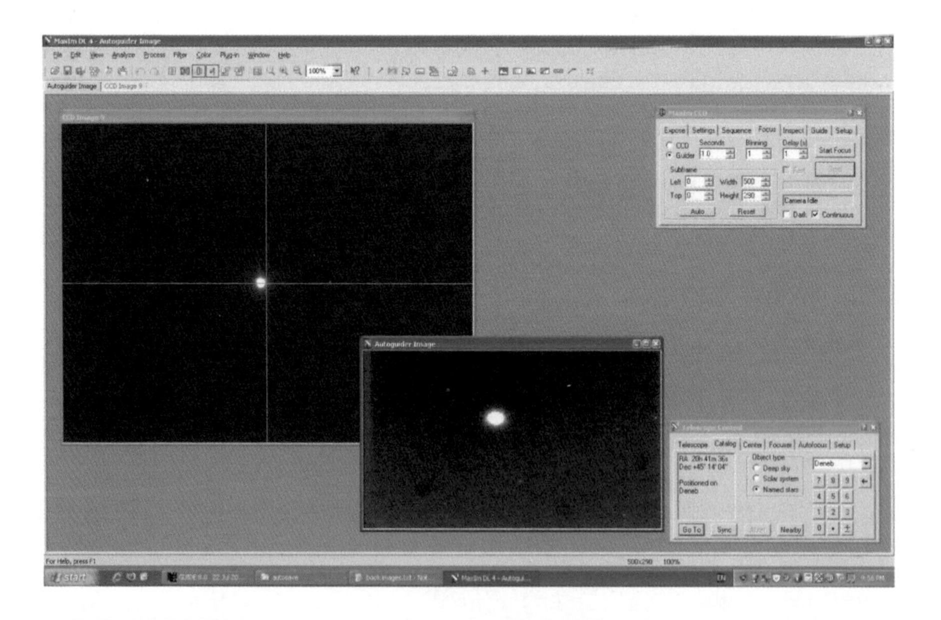

Fig. 7.2 Aligning guide 'scope with main scope. Ideally performed using a bright star

exposure should immediately produce a discernable star on the guide camera window. You can now focus the guide 'scope to obtain the sharpest possible star. For precise focusing, you can reduce the exposure and/or use another field star, ensuring that the star is not saturated (the saturation level is 255 on my guide camera). The *inspect* tab shows both the maximum level and the FWHM (full width at half maximum signal level) of the star. Try to obtain the smallest possible FWHM level. Remember that this exercise is to obtain the best possible focus. The choice of guide star comes later.

Having now focused the guide 'scope, you can drive to your first (test) target. Use the catalog to select a suitable test target. This can be any target that you wish: bright star or deep sky, depending on the season and time of night. Use a well-known target because you know what such targets should look like. When the target is reached, use the main CCD *expose* option to take a sample image. Use a 10-s exposure and then plate solve using the *Pinpoint* module within *Maxlm*. This solving exercise provides confirmation of the current effective focal length (and therefore focal ratio) of the optics, and the angle of rotation of the camera relative to the RA axis. Always take time to adjust this angle to near-zero before the imaging commences. Before trying to solve any image, ensure that you have at least entered approximate values for the focal length into *Maxlm's* telescope details. This process ensures that the correct FITS header information is loaded because the plate will not solve without valid data, although the processing data can be edited prior to the Pinpoint run. Ensure that the scope and guider data are accurate, so that the FITS data is valid.

Figure 7.3 shows that enough stars were detected (order = 2) to calculate the RA and Dec of the center of the image, and noting that the rotation of the camera horizontal axis is +00°38′ relative to the RA axis, and that the effective focal length of the current

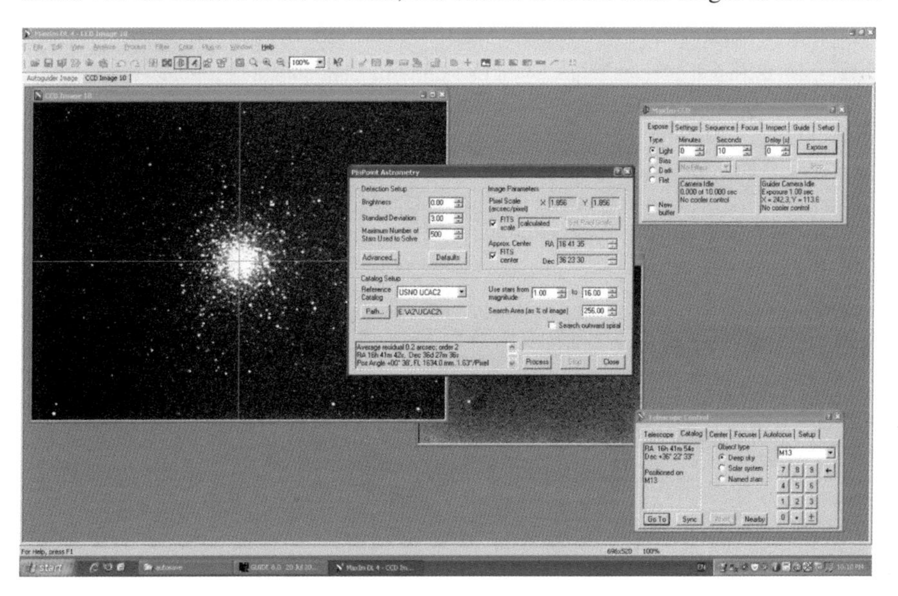

Fig. 7.3 *Maxlm DL* (Pinpoint LE) plate solve from 10-s exposure of M13

optical system is 1,634 mm, giving an effective resolution of 1.63 arc s/pixel. Not bad for one plate solve! If necessary, re-edit the telescope details so that the now accurately known focal length is recorded in the FITS header. This result can be taken further. Having just solved the image, you can select *sync* on the telescope control module. This gives the 'scope a precise position. Then click *goto* on the same module and the scope will gently slew the small distance required to position it exactly on target. In this instance, you can see from the main image that you are already there, but in many cases following a long slew, the scope may be several minutes of arc off target. A plate solve followed by a *sync* and *goto* should then place the target dead center. Clever stuff!

A second solved image of a completely different target should produce virtually identical results for the focal length and other parameters. The imaging setup is now working well. Notice another factor in both images. For effect, the author tried different star catalogs for each image solve. The *USNO-A2.0* catalog was used to solve one image and the *USNO UCAC2* catalog to solve the M13 image. *UCAC2* is a more recent catalog and includes star positions of greater precision; this is why the average residual error is smaller for the result from this catalog. Now on to calibrations!

Guide Star Dark Calibration

Ready to image? Almost! When everything is correctly configured, for the best prospect of obtaining a quality guide star image for guiding, you should use the *Maxlm* guide calibration feature. This process is described in detail in the *Maxlm* manual. Briefly, configure the main camera and guide camera as *swapped heads*, and then take a set of (say) 10 bias images from the guider, and 10 darks (of about 3 s exposure). Using *Maxlm's* calibration feature you can produce dark subtracted exposures from the guide camera for guide star selection. There is one important note here. This feature does not work with the *Starlight Xpress* adaptive optics unit, at least as of versions 4.62 and 5.06. Later, we shall explain how to use varying guide exposures to identify potential guide stars, so a calibration using bias is essential. From this calibration, you can subsequently use any guide exposure and still obtain optimum subtraction.

Calibration of Autoguider

The guide option on the CCD module offers *expose, calibrate*, and *track*. *Expose* allows the guider duration to be adjusted. *Calibrate* causes the guider calibration to proceed. Drive the 'scope to the meridian near zero declination. Use the guider focus option to provide a steady stream of short exposures that allow you to see stars. Drive the telescope slowly until a field is found that shows a reasonably bright guide star and preferably no nearby bright stars in the same field.

When you have found one, terminate the focusing and take a 1-s exposure of the region using *expose*. If you have selected a good field, the bright star will be selected unambiguously. Ensure that it is not saturated, as indicated by the guide star display. For future reference, note that you can force *Maxlm*'s star selection window to use a different star by merely clicking the mouse on the alternative star. If all is well with the brightness of the selected star, slowly drive with the *focus* option restarted, until the measurement (calibration) star is nearly central in the field of view; drive in both RA and declination as necessary.

Under the *guide options* tab, select *guider settings* and set the settings for the X axis and Y axis times to (for example) 10 s. This is the time that the scope will be driven (at sidereal rate) forwards in RA and then declination, in order to produce changed positions of the guide star in the autoguide camera images. These distances will then be used to calculate the amount of movement correction required to control the telescope during autoguiding. The scope is then driven in RA by the calibration software towards one side of the field of view.

If the movement is less than five pixels, the process fails because the calibration requires a reasonable movement for this measurement. If this should happen, increase these (X and Y) settings. Similarly, if the guide star moves too far, reduce the settings. Aim to produce a total movement of perhaps one quarter to half the field of view. After the second image is displayed, the scope is driven back to the first position and a third image is taken. Ideally, no backlash should be seen, so the second image should see the guide star back where it started. A line drawn on the screen shows the travel. The scope is now driven in declination and a further image is taken; this should be at 90° to the RA line. At the end of the calibration, the guide star should have returned to its original position (see Fig. 7.4). There may be an indication here of declination backlash.

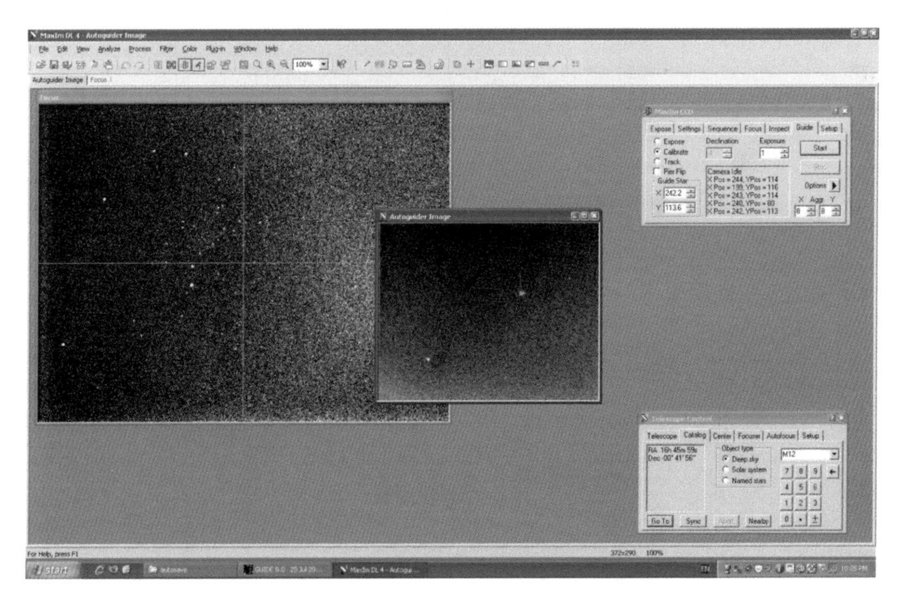

Fig. 7.4 Maxim guider calibration completed, showing the *red lines* at 90° and indicating success

As can be seen from Fig. 7.4 the guider calibration has been completed successfully, shown by the clean right angle that records the movement of the measurement star. The upper right window shows the individual positions recorded during the telescope movements. If your guide camera is going to be used regularly – or even if not – this is a good opportunity to orientate the guider, so that the axes are aligned with RA and declination. After making this adjustment, repeat the calibration to ensure that the latest results are valid.

If this calibration fails, for example when the lines are not at 90°, the most common cause is a second bright star appearing in the field and confusing the software. If this happens, find a new, lone bright star. Careful exposure adjustment will often rectify the problem of software guide star confusion. A study of the *MaxIm* manual will provide further understanding of all the parameter adjustments available for autoguider operations.

Autoguiding Test

Once you have calibrated the autoguider, you can try your first autoguiding session. For this, you need to find a suitable target, perhaps a deep sky object reasonably high in the south. With optimized autoguiding settings, the sky is your oyster. But before getting underway with serious imaging projects, it is important to ensure that your settings are optimized for your telescope combination.

Select your target, center it in the main camera prior to the long exposure, and ensure that focus is as good as it can be. You can focus using a Hartman mask. Also try to remember to remove the Hartman mask after focusing! When the main imager is ready, take a sample autoguider image (use *guide, expose*) for 4 s, and examine it. Focus should be perfect – or nearly so – in order to ensure that sufficient guide stars are potentially available. If all is well, the software will identify a suitable guide star and display its coordinates and brightness in the guide window.

As described earlier, longer exposure guide images are less likely to deliver erroneous corrections, so a 3–4 s exposure is potentially an ideal solution. Select *track, options*, and *tracking error graph*. You can customize the type of plot that you want; you might want to use a line plot with a vertical scale of two pixels. Change the X and Y *aggressiveness* settings to 8 and 0, respectively (declination correction disabled); select *start* and watch the autoguiding process happen. It can be very revealing!

Figure 7.5 shows a fairly typical tracking plot using a 4-s guide exposure with a bright guide star. The picture shows that the second exposure is under way (201 s so far) and the guiding has been very good, showing minimal variations during the whole period. The upper plot is RA error and because the periodic error has been minimized, there have been few corrections required. This target is fairly high, so that would also contribute to a good result. To show the consistency of this guiding, Fig. 7.6 shows the same guiding result on a different target.

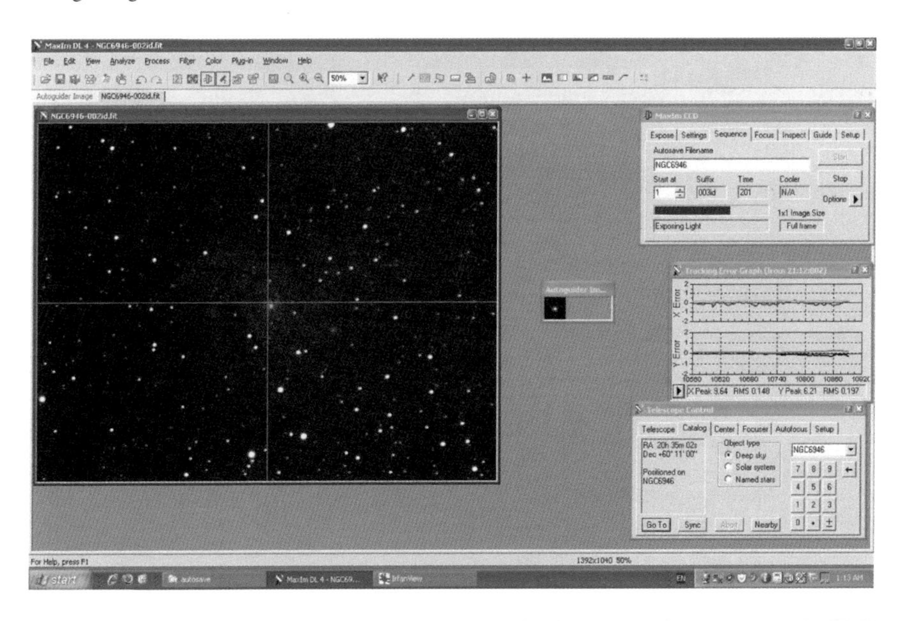

Fig. 7.5 Tracking plot from guiding on NGC6946 using 4-s guide exposures

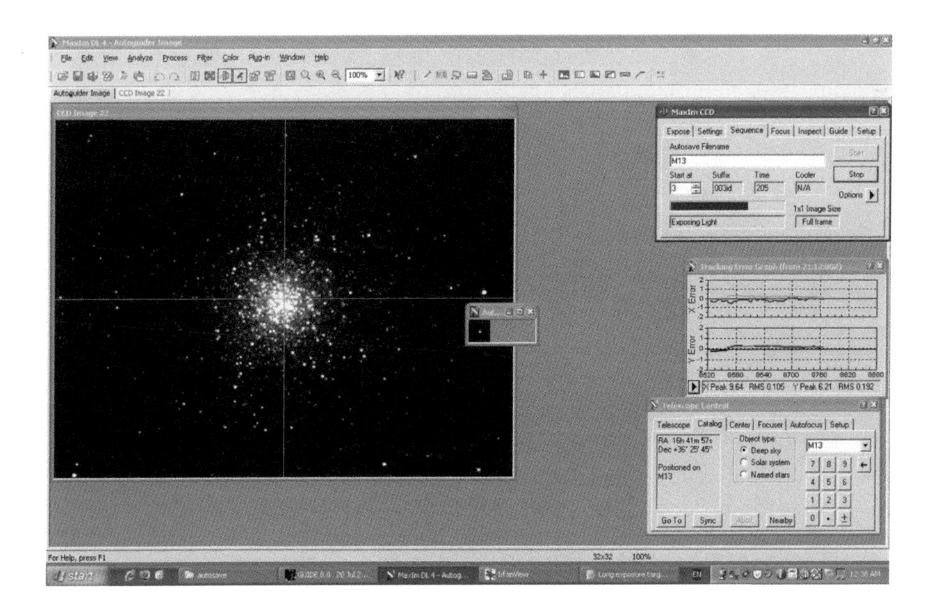

Fig. 7.6 Guiding on M13 using 4-s guide exposures

Fig. 7.7 Final image of M13 using 5-min exposure with IDAS filter

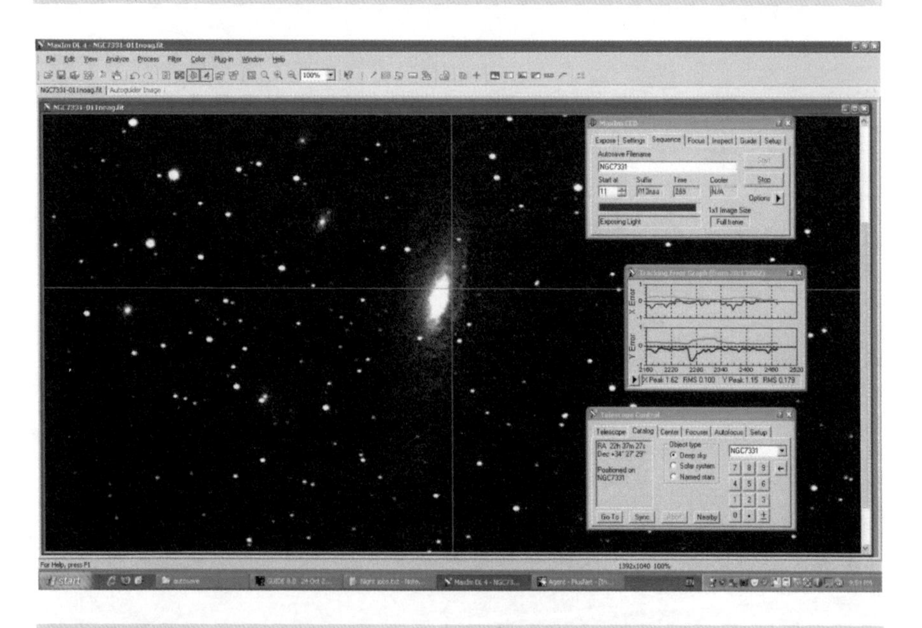

Fig. 7.8 Entering and leaving the observatory – a kinky event

In most cases, you can expect a well-guided result, as seen in Fig. 7.7, which shows a newly downloaded full-resolution raw SXV-H9C image of M13. Close examination of the actual screen dump shows almost perfectly round stars. For guide 'scope autoguiding imaging, if necessary use an IDAS (light pollution suppressing) filter.

Mind Where You Walk!

What Follows is a Cautionary Tale!

During the testing phase of my autoguiding setup, I was monitoring guiding from my home via my network when I began to notice occasional kinks in the declination tracking error plot – see Fig. 7.8 – that aroused my curiosity. After seeing this kink repeating every few minutes, I realized its cause. Every time I entered or left the observatory, the declination axis suffered a movement of about three quarters of a pixel; this took a few seconds to correct. Fortunately, the effect on the final image was virtually unnoticeable! It does illustrate the sensitive nature of the process of autoguiding.

References

Jim McMillan's article on Autoguiding
http://www.stargazing.net/David/astrophoto/AstroPDF.html

Chapter 8

Previous chapters have discussed the setting up of your Meade LX series telescope, described the processes of polar alignment, periodic error measurement, and correction, and also described processes such as Smart Mount calibration that are built-in to the scope to aid operations. All these processes are designed to raise your telescope to peak performance. Comparable processes can also be applied to many of the higher cost telescopes on the market today. Chapter 9 describes one other piece of equipment – an adaptive optics unit that can take you about as close to the limit of what might be possible as you can get. This chapter is about additional software that can greatly improve the efficiency of imaging sessions.

As we have seen, the LX-series of telescopes does come with software (the *AutoStar Suite*) that is adequate to control the scope for routine visual use, as well as provide some imaging applications. Meade also markets CCD cameras, and these products include appropriate imaging software. If you read the forums devoted to these telescopes you may notice occasional references to research projects being undertaken by enthusiastic amateurs. Such project descriptions often include references to software with which you may not be instantly familiar. The Meade LX series scopes – particularly the LX200GPS and LX400 models – are fully capable of enabling astronomers to pursue serious astronomical projects, as described in Chap. 11. That is why so many individuals and groups, amateur and professional, use these telescopes. Such projects invariably require advanced software that has been developed in recent years.

During early efforts to improve image quality and the efficiency of imaging sessions, the author wanted to pursue such projects that almost represented a return to professional work. A number of software writers had produced scripting software able to command the telescope through a series of individual targets throughout an observing session. In this chapter we are looking at sections of

L. Harris, *So You Want a Meade LX Telescope!*, Patrick Moore's Practical Astronomy Series, DOI 10.1007/978-1-4419-1775-1_8, © Springer Science+Business Media, LLC 2010

some highly capable programs that can produce excellent results with an optimized telescope.

A Day (Night?) in the Life

An astronomer working on a scientific project wants to be able to power his or her telescope up, check the focus, have it confirm synchronization with the sky, and take any calibration images that might be required for later processing. He or she would then proceed to the first in a prepared list of targets (a plan), take a sample image, analyze it for positional accuracy, make any necessary position adjustments, ... you get the idea!

The sequence of events is fairly obvious and reasonably well defined; we usually do this almost without thinking as we proceed through a session's imaging. You take the telescope to a succession of targets and ensure that each one is accurately centered (or at least within an acceptable distance). You then guide while imaging for a defined exposure time. In the case of asteroid measurement and detection, you want the scope to drive to the first asteroid field, check using a sample image that the scope is accurately positioned, and then take your relatively short exposure image without autoguiding. If the next asteroid field is a significant distance from the previous asteroid, you want the scope (or at least the software) to check that it has arrived at the correct position before starting imaging. You can define suitable numbers (such as acceptable distance errors from the target) in order to have the scope do exactly what you wish at the right time. It is not difficult therefore to compile a plan that lists the individual actions required.

Serious imagers want to go *much* further than simply moving from target to target; an imaging session typically sees the scope sitting on a small number of different targets for extended periods of time while the telescope is carefully autoguided on each and the camera collects multiple images for later combining. A set of filters might also need to be sequentially selected if a color image is required. Finally, you want to park the telescope safely and even close the shutter if you have an automated observatory.

This summarizes most if not all of the tasks that a busy astronomer might wish to do during and after an observing/imaging session. The programs discussed in this chapter can be used to perform some or all of the tasks – to a greater or lesser extent. The simplest of them – *Astrometrica* – will be examined first, leading to more advanced programs, including those able to take complete control of the observatory for the night.

Astrometrica

Astrometrica is a low-cost software tool written by *Herbert Raab* (see Fig. 8.1) that can be used for the scientific astrometric data reduction of CCD images.

Fig. 8.1 Herbert Raab

Astrometric data reduction is the science of measuring the precise position of heavenly bodies. To measure a position one could use meridian circles (these produce absolute positions) or reference stars (producing relative positions). With modern computers and large databases of accurate star positions available, *Astrometrica* uses reference stars.

Software only has to identify a dozen of these reference stars to produce an accurate position. Such matching immediately correlates measured magnitudes and known magnitudes, consequently deriving error differences and also positional differences. Also as a result of this one-to-one matching, any detections not identified can be highlighted for potential further investigation. If you provide two similar exposures of the same field (taken at different times) you have the potential for asteroid and supernovae discoveries!

Meanwhile, back to *Astrometrica*. The latest version for the *Windows* operating system originated as DOS-based software that was first used for the astrometric data reduction of photographic films, and later, CCDs. Its main features include the following:

- Reading FITS (8, 16, and 32-bit integer) files. Image size is only limited by the available memory.
- Automatic image calibration (using dark frame and flat field correction).
- Blinking with automatic image alignment.
- Zoom and magnifying glass for close-up image inspection.
- Automatic reference star identification.
- Automatic moving object detection and identification.
- Track and stack function to follow fast or very faint moving objects.

- Access to the complete MPC database of orbital elements.
- Access to new-generation star catalogs (USNO-B1.0 and UCAC 2).
- Internet access (sending e-mail data to the MPC; downloading the *MPCOrb* database.)
- Online help system and tutorials.

If you are proposing to install the software for the first time (not just upgrading it) you should download the complete package – approximately 5 Mb. At a reasonable cost, with an extended period permitted for trying it out, this represents extremely good value for money considering its capabilities. The normal application for this program is the detection and identification of asteroids, but it has other useful features.

Before attempting to solve the first image, various parameters must be entered in the *file/settings* set of tabs. Details of your observing site should be entered, then your equipment details (telescope and CCD) on the *CCD* tab. The *program* details for entry include the catalog path data and object detection specifications. Entry of the *environment* details should largely complete the essential requirements. The *Internet* tab provides the ability to send an e-mail to the Minor Planet Center (MPC) when you are fully operational. The software is tolerant of moderate parameter errors – such as focal length and camera orientation – when calculating image properties.

Follow any change of optical configuration (e.g., f8 to f6) by entering nominal values for these data into both the image control reception software (*MaxIm DL*) and also in *Astrometrica* to ensure that the FITS headers are correct. This leads to a probable solving of the image. If these parameters are left unedited in the original camera control program, the subsequently collected images may not solve easily because the image scale is likely to be too far out, and in any case, could contain erroneous header information.

One application for this program is its field star analysis feature. After taking a first nominal exposure image (say, a 60-s exposure) of a star field, you can use *Astrometrica* to analyze the field. This takes a few seconds, after which it presents a modified view of the image showing the majority of the stars having color-coded circles around them, indicating that the software has uniquely identified each one from the reference star catalog.

The sequence is as follows: after entering the settings, the image for solving is loaded by selecting either *file/load* or clicking the relevant icon. When loaded, select the *Astrometric Data Reduction* icon, and after confirming that the displayed data is correct (as taken from the FITS header – hence the importance of accurate entry) the image is analyzed and hopefully solved. If it fails to solve, you have to follow the on-screen instructions and adjust the settings. It is unlikely to fail if you have entered reasonably good data, but occasional problems can cause solution failure.

List 8.1 shows the results from solving a typical image. The Data Reduction Results table is displayed within the solved image, showing the number of stars detected and the number of stars positively identified by comparison with the star catalog previously referenced under the *program* tab. Several other results are also

List 8.1 *Astrometrica's* star analysis

16:24:39 – Start 2008/12/13
Image 1: C:\autosave\20081206\NGC7380-001f6aoid.fit
Timestamp: 2008 12 06, 18:31:50 UT

Settings for scale and orientation
Focal length = 1,430.0 mm ± 1.0%, Position angle = 0.0° ± 10.0°, Pointing = ±5.0′
Image flipped: No

Settings for CCD
Pixel width = 6.5 μm, Pixel height = 6.5 μm, Saturation = 63,000

Settings for object detection
Aperture radius = 3, Detection limit = 2.5, Min.FWHM = 0.70, PSF-Fit RMS = 0.20, Search radius = 0.75

Settings for reference star matching
Number of stars = 50, Search radius = 2.0, Magnitude = 8.0–18.5 mag
16:24:45 – USNO-A2.0: 4,481 Records read (31.7′ × 26.2′)
Center coordinates: RA = 22 h 47 m 19.00 s, De = +58° 08′ 03.0″
16:24:45 – object list for image 1 (NGC7380-001f6aoid.fit):
1,464 detections (871 Stars, 871 Ref. Stars, 0 Movers)
16:24:45 – Astrometry of image 1 (NGC7380-001f6aoid.fit):
830 of 871 Reference stars used: dRA = 0.27″, dDe = 0.27″
$X = -1.734455580\text{E-}5 + 4.491585639\text{E-}6 * x' + 4.970785855\text{E-}8 * y'$
$Y = +3.613320040\text{E-}5 + 4.944801203\text{E-}8 * x' - 4.492544042\text{E-}6 * y'$
Origin: $x_0 = 696.0$, $y_0 = 520.0$
Center coordinates: RA = 22 h 47 m 19.45 s, De = +58° 08′ 10.5″
Focal length = 1,446.9 mm, Rotation = 0.63°
Pixel size: 0.93″ × 0.93″, Field of view: 21.5′ × 16.1′
16:24:45 – Photometry of image 1 (NGC7380-001f6aoid.fit):
582 of 871 Reference stars used: dmag = 0.44 mag
Zero point: 26.86 mag

displayed here. Select *file, view log file*, and a complete analysis of the identified star field and parameter list is displayed (see List 8.1). This lists the parameters previously entered into the *settings* tab for confirmation, the catalog used for analysis, the plate coordinates and center, the object list with a detection summary, and the measured parameters of the telescope and camera pixel dimensions.

The latter parameters – focal length and camera orientation – are valuable data that can be immediately used. You can now enter the accurate focal length into the camera imaging program's FITS entry point so that all subsequent images are correctly labeled. You then mechanically adjust the camera orientation and take another sample image – usually of about 10-s duration. This exposure is normally long enough for a sufficient number of stars to be detected to give a valid orientation analysis. By repeated orientation adjustments, the orientation can be brought to within 1°. You are then ready to complete the rest of the imaging session. Maintaining an orientation as close to zero as possible means that all images will at least align well and have maximum overlap. Subsequent images of star fields taken for asteroid and particularly supernova hunting should subsequently display comparable star fields.

Further down in the log file is specific star data of which Table 8.1 is a small selection from an image analysis of NGC7380. Apart from Right Ascension and declination data, the magnitude and SNR (signal to noise ratio) data is invaluable. Scanning down the list of star magnitudes instantly shows how good the seeing and focus are.

On this particular night a northern airstream had provided very cold but clear skies in which magnitude 18 features several times in the list (of which one entry is seen above). On most nights you might hope to see several magnitude 17 reference stars listed, so this was a good night – at least during the early stages before mist and fog terminated the session. The whole process of running *Astrometrica* and analyzing the first image takes about 1 min and delivers sky magnitude limits. Experience suggests that if the software cannot see below magnitude 16, preferably magnitude 17, then the skies are not currently very clear. At least you can then make an informed decision about the nature of the current session's possibilities. Regional weather forecasts and home weather satellite monitoring (see Chap. 10) provide a guide for every session.

Also shown is the FWHM (full width half-maximum) analysis, indicating how good the focus/seeing is. It helps to perform this analysis routinely every couple of hours or so during the night in order to monitor sky clarity (minimum detected magnitudes) and focus variations. If a 10-min exposure (those used for deep sky target imaging, not astrometry) only gets down to magnitude 16, it can be time to make a decision about whether to continue the session or not.

Asteroid, Comet and Supernova Detection

This is one of the main applications for *Astrometrica*. As described in Chap. 11 (Example Projects) the Meade LX200GPS/LX400 scopes are ideally suited to both asteroid and comet detection and are widely used for these purposes. Essentially the astronomer searching for asteroids or comets will take three images of the same region of sky, separated by perhaps 30 min and exposed for perhaps 60 s each. Depending on the actual target, these periods may vary. For instance, near Earth objects (NEOs) are often imaged for only a few seconds and the images then analyzed, possibly even stacked (added together by software). During these short exposure periods the mount is tracking on its own; there is no autoguiding. This is one reason why it is essential that the scope be properly polar aligned and be operated with minimal periodic error and drive backlash.

Astrometrica can display a group of three time-separated images and then analyze them. After analysis, any object that has been detected moving across the three plates will be highlighted for specific analysis (see Fig. 8.2). A blink feature is naturally included, and this will usually help the observer positively identify a genuine asteroid.

Table 8.1 Astrometrica's analysis of star field around NGC7380

RA	dRA	Dec.	dDec	*R*	d*R*	*x*	*y*	Flux	FWHM	Peak	Fit
h m s	″	° ′ ″	″	mag	mag	–	–	ADU	″	SNR	RMS
22 45 58.051	(+1.27)	+58 06 08.93	(–0.69)	16.45	(+1.45)	1,390.88	657.06	14,613	2.2	18.1	0.100
22 45 58.438	–0.05	+58 12 32.46	–0.11	17.79	+0.79	1,390.09	243.18	4,246	1.2	6.3	0.136
22 45 58.503	–0.27	+58 13 20.69	–0.10	17.68	(+1.38)	1,389.85	191.14	4,708	1.2	5.7	0.130
22 45 58.521	+0.95	+58 14 09.19	–0.36	18.54	(+1.74)	1,390.02	138.80	2,126	1.0	4.5	0.192
22 45 58.533	–0.45	+58 08 22.58	+0.11	17.82	(+1.12)	1,387.64	512.82	4,116	1.1	4.8	0.138
22 45 58.824	–0.15	+58 00 46.00	+0.52	16.16	+0.06	1,382.14	1,005.48	19,006	1.9	15.6	0.085
22 45 59.302	+0.13	+58 05 23.20	–0.15	16.20	–0.30	1,379.88	706.35	18,345	2.6	7.7	0.151
22 45 59.345	–0.14	+58 02 59.79	–0.30	16.78	+0.48	1,378.55	861.08	10,822	2.2	6.8	0.154
22 45 59.361	+0.13	+58 11 14.21	+0.06	13.90	+0.20	1,381.69	327.57	152,214	2.8	66.0	0.132
22 45 59.504	+0.41	+58 10 56.08	+0.24	17.27	+0.77	1,380.35	347.13	6,852	1.2	8.5	0.090
22 45 59.671	(+1.64)	+58 08 50.09	(+0.28)	17.32	+0.72	1,378.09	483.08	6,550	1.2	8.8	0.096

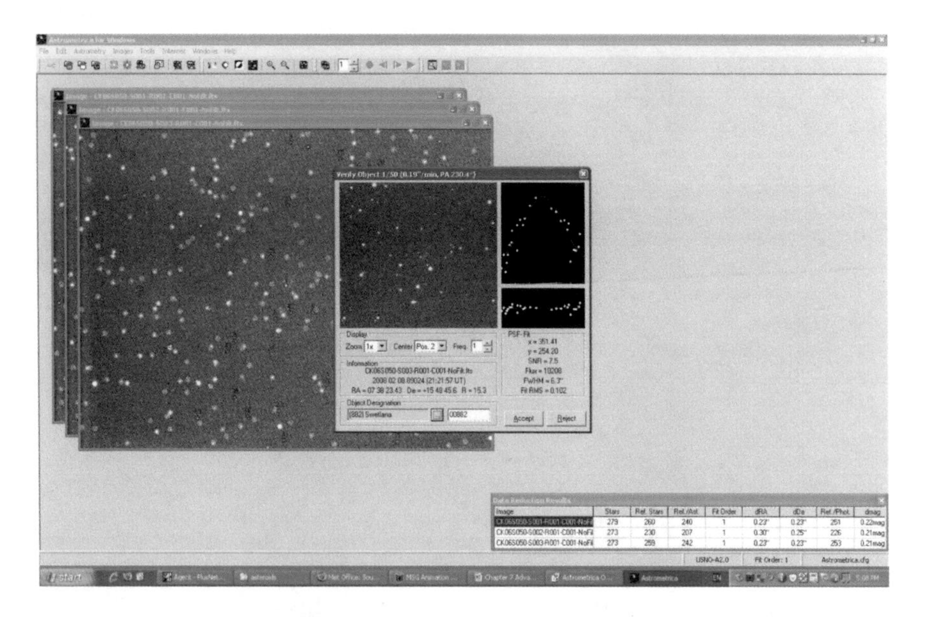

Fig. 8.2 *Astrometrica* detects and displays an asteroid in three consecutive images of asteroid 882

Analysis can be done in various ways. The images can simply be blinked using the *blink current images* icon, as described earlier, and can also be fully analyzed using the *moving objects detection* icon. Figure 8.2 shows asteroid 882 (Svetlana) detected in the three images. All detections, whether stars or asteroids, are color-coded, and the color can be identified via the screen icon and changed to your preferences. The included tutorial explains how to obtain actual astrometry for each asteroid detection.

At least during the early stages of use, the software is likely to make a number of false detections. By reading the notes about parameter optimizing you should be able to gradually tune the settings to minimize the number of false detections. Sometimes a burst of noise within the image may be misidentified as a possible asteroid. This is indicated within the blinking window, where it tells you which detection you are viewing and asks for confirmation or rejection. It is often better for you to make a personal visual identification in cases of software-detected ambiguities. If a large number of misidentifications are made it *suggests* that the detection parameters need to be adjusted. *Astrometrica* also offers the *track and stack* feature where a number of images taken in close sequence can be combined. If your project is supernova hunting, then blinking two images – one being a reference image taken some time back and the other being the latest image of the same region – should show any new object.

To summarize, *Astrometrica* is a low-cost program that is fully capable of providing a very quick evaluation of sky and telescope conditions to help judge the potential of a session's observing, as well as being an efficient astrometric analysis tool for the detection of asteroids, comets, and supernovas.

The program can be downloaded from the following site: http://www.astrometrica.at/

CCD-Inspector

This program comes from the same stable as *PEMPro*. *CCD-Inspector* is an image quality analysis program written by Paul Kanevsky. It analyzes previously obtained images and reports on their optical quality, and – more importantly – it can be used in real time for optical adjustment assessment and improvement. This means that you can assess and tune your optical system's performance with real-time optical analysis technology.

CCD-Inspector provides 3-D field curvature maps, visual FWHM displays, and a detailed analysis of the quality of FITS images. The availability of a 30-day trial period made it possible for the author to analyze several recently acquired images and to analyze and improve the current optics of his telescope – specifically the collimation. It was not possible to complete the tests due to an unprecedented long period of cloudy weather over Britain. Even a nominal 30-day extension did not add materially to the experience, providing only two short further sessions. It was, however, possible to use and appreciate the real-time collimation feature.

As established in Chap. 2, the advanced Meade telescopes are designed to provide adjustable collimation using – in the case of LX200GPS telescopes – three screws at the front end of the scope in the central part of the secondary mirror. This adjustment was described in Chap. 2. LX400 scopes have built-in electronic collimation adjustment controlled on the keypad, one of the scope's various advanced facilities. It was therefore easy to try out the program.

After starting the program look at previously obtained images to see how *CCD-Inspector* analyzed them. Using the conventional *file: open* you can select one or multiple files to load. These can then be measured as a group and the results displayed. The program immediately shows how far off the collimation is! Most interestingly during the author's tests it also indicated a significant tilt. A very close look at the scope's optical system, particularly the interfaces, revealed that one of the adapters within the adaptive optics unit and camera section was very slightly out of alignment. The adapter rings are held in place by two tiny screws, and one was slightly loose, allowing a small (but significant) misalignment to be introduced. This was easily corrected. Clearly the correction would only affect new images.

The analysis of an early test image displayed the results of analyzing 189 stars and showed that the overall tilt was 19°, with a collimation error of almost 11 s of arc. Some of the other images taken at that time showed comparable errors, though not identical. Following the later expiry of the trial software it was impossible to remeasure any older images.

In the author's experience familiarity with the use of the software was a great benefit. While investigating the saving of the new collimation state of his LX400 he tested an aspect of the focus feature without yet having saved the collimation status. Unfortunately, this process caused the motors to run full length to the end of

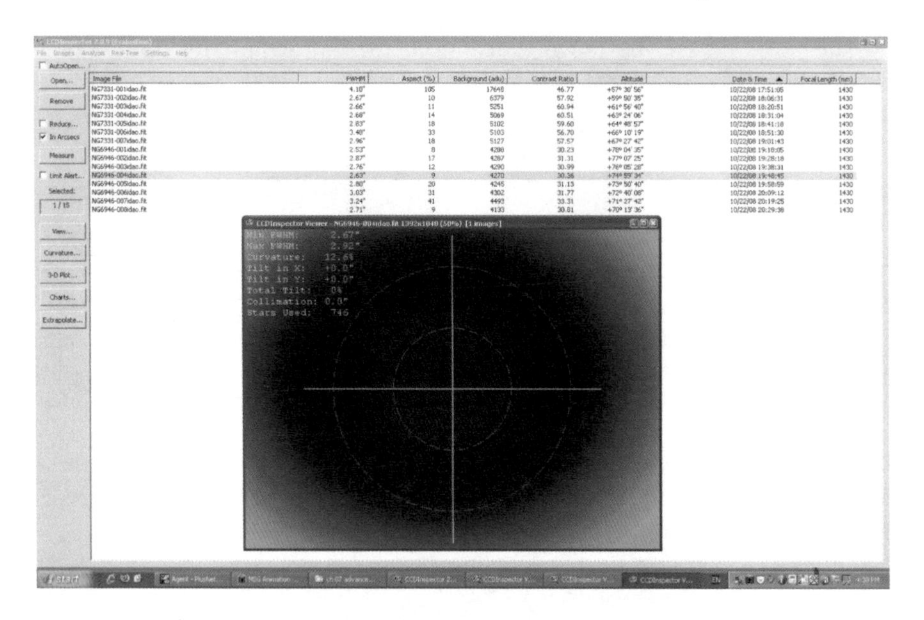

Fig. 8.3 Collimation group analysis after adjustment showing great improvements

their tracks before retracing their steps. Having then realized the omission it was not difficult to repeat the entire collimation process using the software, this time completing it in a much shorter session. Figure 8.3 shows the much improved collimation and tilt.

The process of measuring collimation parameters in real time can be done in two entirely different ways with *CCD-Inspector*.

1. Measuring in real time: In this method you expose and download an image (it can be binned), and the software derives the collimation parameters. This process is not difficult, though you do need to expose images for several seconds in order for the software to obtain enough valid stars for measurement. If there are not enough stars, the software reports this as an error message.

   To start the collimation process you must select a suitable star field. By looking at a number of high-elevation NGC targets on a planetarium program such as *Guide-8* you find providing an adequate number of stars an adequate number of stars over the camera's field of view. Insufficient stars across the field will produce an error message. It actually required a 20-s binned exposure to produce a reliable measurement of collimation parameters. A measurement using 362 stars derived an error of 10.7 arcsec. As collimation improved, the error measurement slowly decreased, finally showing a considerable reduction, down to below 2 arcsec (see Fig. 8.4).
2. Single defocused star collimation: This innovative method uses a single bright defocused star for measuring and adjusting collimation. You select a high-altitude bright star – for example, Scheat – and then defocus it to obtain a fairly

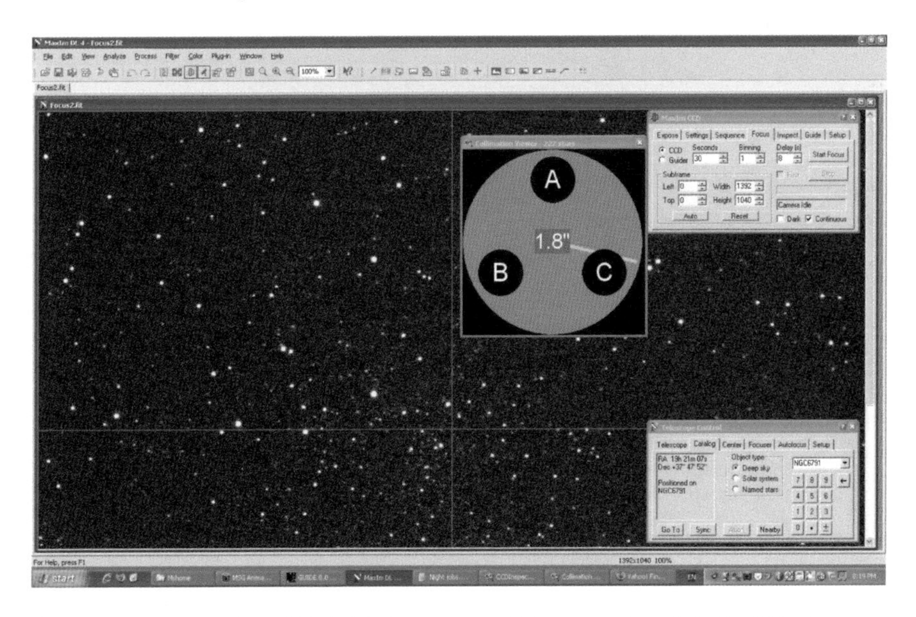

Fig. 8.4 Real-time collimation significantly improved

large circle. The star must be centered within the image field and exposed for long enough to ensure a good signal-to-noise ratio. You can use an initial exposure of 2 s, with *MaxIm DL* running in *focus* mode producing a sequence of images. You can later reduce this to 1 s; it is important to not reduce it below 0.5 s to avoid monitoring changes in seeing conditions. *CCD-Inspector* is then run, and the acquisition process started with *Settings, camera control software*. The defocused images can be binned, so you can start this way. You do not need to download a full image; a subframe should be enough. The collimation viewer is then started (*real-time, collimation, defocused star collimation viewer*) and then the image viewer (*real-time, image viewer*).

It is possible that you may see a display showing the defocused star with the image being refreshed regularly, but with the one-star collimation viewer registering 0.0 (indicating measurement is invalid), and the image viewer apparently empty. By right-clicking the icon at the top left of the image viewer, the display settings can be changed, bringing the star on to the display and activating the measurement.

Changing the image viewer screen display parameters quickly reveals the measurement star. The adjustment process is adequately described in the online help instructions. The defocused image can be adjusted to occupy most of the circle. As you watch the sequence of measurements you can see how far off collimation the current optics are. The aim now is to adjust the collimation so that the star moves in the direction of the arrow. This should be done carefully while pausing to check that the direction is correct. Success should be indicated by a gradual reduction

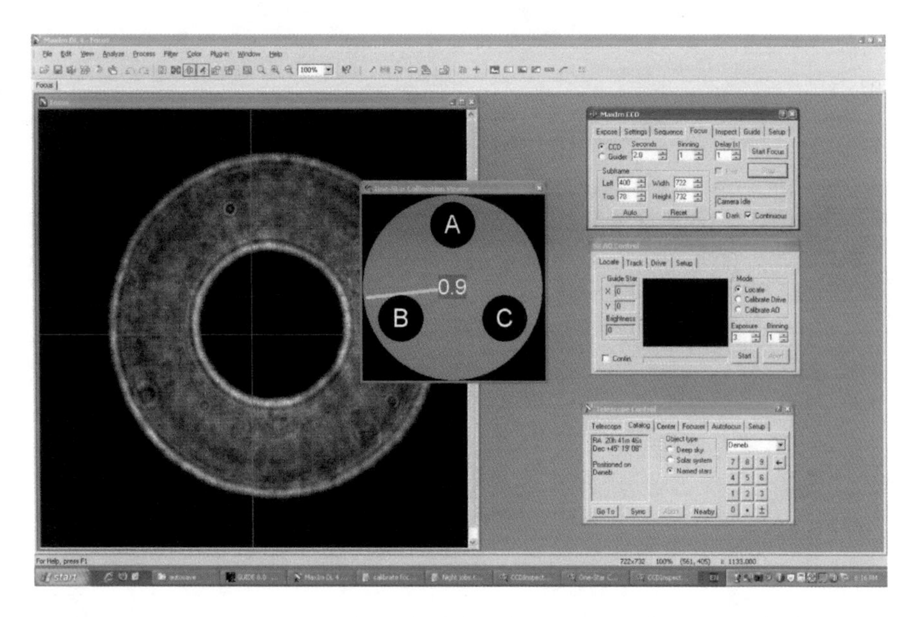

Fig. 8.5 Improving collimation using full-resolution imaging

in the measured and displayed error. You need to carefully move the scope to bring the image back within the boundary during the process, and again watch for the measurements to stabilize. When all is working well, you should find that you can reduce the collimation error in a systematic manner. After obtaining an improved collimation, you can change to full resolution (from previously binning) – see Fig. 8.5.

By the time that errors are small, you should reduce the amount of adjustment. Eventually you should see the arrows changing direction and magnitude as you approach the limit of what is possible under your current seeing conditions. The instructions for the LX200GPS process remind you of the importance of not repeatedly turning one knob in a counterclockwise direction, which would eventually loosen the secondary mirror.

CCD-Inspector can be a useful program for measuring and improving the optical collimation of the scope. Improving the collimation of your LX200GPS or LX400 scope ensures that your optical system will give its best performance in terms of deepest magnitude and best spatial resolution. The real-time measurement using a bright defocused star was particularly convenient and effective.

Download CCDInspector2 from http://www.ccdware.com/

PoleAlignMax/FocusMax

Both these programs were written by Larry Weber and Steve Brady and are issued as freeware. The programs are *ASCOM*-compatible and require either *Maxim DL* (from Cyanogen) or *CCDSoft* (from Software Bisque) in order to be used. The first program measures star field orientation in two regions across the sky and deduces the error from true polar alignment. After adjustments, this process can be repeated. Full details of the latest instrument control capability, and download instructions can be obtained on the Web site http://users.bsdwebsolutions.com/~larryweber/.

AstroArt

This program is a complete software suite for image processing, photometry, astrometry, CCD control, and image stacking for CCD and film images. All major CCD cameras, webcams, and DSLRs are supported by *AstroArt*. Focusing, autoguiding, imaging, and scripting are easy and quick.

AstroArt is also an open system with plug-in support. All programmers of C/C++, Visual Basic, and Delphi can create new filters using the free development kit. Some of the main features of *AstroArt* are as follows:

- *Advanced filters*. Maximum entropy deconvolution, Richardson Lucy, adaptive, Larson Sekanina, unsharp mask, deblooming, DDP, FFT, and with full image preview.
- *Astrometry and photometry*. with integrated star atlas, GSC included, USNO and UCAC2 supported. MPC reports, aperture photometry, estimation of measuring errors. MPCORB minor planets.
- *Image stacking*. Automatic on sets of images, with subpixel precision. Six algorithms (one star, two stars, all stars, planet, correlation, star pattern), drizzle, color synthesis, and batch processing.
- *Color imaging*. Native 96 bit, visualized in real time. All filters work on color images. Quadrichromy LRGB and WCMY with automatic L.A.B. conversion.
- *High speed*. *AstroArt* is an extremely fast software – thanks to its code optimized in assembly. Fully compatible from *Windows 98* to *Windows Vista*.

There is a demo version available; for more information about the program, visit the Web site http://www.msb-astroart.com/.

The software costs approximately $185 plus transport and comes with tutorials and a number of useful accessories.

MaxIm DL

Douglas George – see Fig. 8.6 – is a leading-light in the Cyanogen stable of software. We have provided descriptions of some aspects of this program in Chap. 4 because it offers effective control of both telescope and camera equipment as well as being ASCOM-compatible. The program also provides more advanced features that merit mention in this chapter.

Here is a brief summary of *Maxim DL*'s V*ersion 5* features as supplied by Cyanogen:

Fig. 8.6 Douglas George

Complete observatory integration. Control of your camera, autoguider, filter wheels, focusers, camera rotator, telescope mount, and dome – when fitted. Built-in miniplanetarium with *AllSky* and *Zoom* view to enable you to see a whole sky overview, with the telescope position overlaid. Large object database and powerful search capabilities.

Camera controls for faster and easier imaging. Create your own presets for finding, centering, focusing, and LRGB sequences.

Multiple auto-dark buffers allow you to quickly switch between presets without reshooting darks.

Quickly change program configurations. If you use multiple camera/telescope setups, you can instantly reconfigure all software settings.

Automatically identify stack groups by object ID and color filter. Stack multiple images with hundreds of individual subexposures. Automatically reject poor quality images based on criteria you choose.

Align images using a variety of automatic and manual alignment techniques.

Color combines RGB and LRGB image sets, using a single alignment reference image.

Includes a complete image processing toolkit, including a large variety of filters, deblooming, curves, deconvolution, color tools, and more.

MaxIm DL works with third party products such as *ACP Observatory Control Software* (see later in this chapter), *CCDAutoPilot 3*, and *PEMPro* to control both telescope and camera.

Apart from the real-time command features for telescope and camera, there are various options for processing the resulting images. Several years back the author was producing sequences of six 10-min one-shot-color (OSC) images (therefore totaling 1 h). He would simply convert them to color and then add them. This would often leave saturated image areas and series of hot pixels, both contributing to spoiling the images. After spending some time researching image processing in the *MaxIm DL* online tutorial it was realized that by simply invoking the *dither* routine (moving the true image center by a few random pixels before each image is collected), subsequent images could be combined using *MaxIm DL*'s *SD mask* or *Sigma clip* routines. This would simultaneously eliminate most of the hot pixels and leave even the brightest parts of the images at their optimum values. Many other advanced processing techniques such as specialist filters, deconvolution, and drizzle are available.

Photometry and astrometry are fully catered for. Photometric measurements of stars are performed using a variety of filters including Johnson UBVRI, Kron/Cousins UBVRI, and Bessell. Comparing different filter bands allows the determination of a color index for a star. *MaxIm DL* includes two photometric measurement tools based on aperture photometry, measuring the total light within a circular aperture. The simplest is the *Information* window, which can be calibrated to a star of known magnitude on an image. Then other stars can be measured using that calibration. Cyanogen offers a 30-day trial of *MaxIm DL*, and the various options for purchase can be seen on their Web site (select "store"): http://www.cyanogen.com/

ACP Observatory Control Software Suite

The Meade LX200GPS and LX400 series scopes come with built-in features (accessed by the *Autostar II* handbox) that enable you to run your own tours of the sky. A description of how to prepare such tours is given in the manual for each scope. It involves preparing a simple text script of your selected targets. You can also download prepared tours and other data from Meade's Web site. If you used the *AutoStar Suite* software during the early stages of configuring the scope, and possibly while performing the Smart Mount (SMT) pointing routine, you may have seen references to these prepared tours. In this section we are looking at the *ACP* program suite because it offers a complete solution for the astronomer wanting to have full computer control of his or her observatory equipment. *ACP* is from DC-3 Dreams, a company run by Bob Denny (see Fig. 8.7).

ACP Observatory Control Program includes many separate features and advanced functions, as well as other associated programs. In this review we shall look at the three main components: *ACP*, *ACP Planner* software (that actually comes free), and the full version of *Pinpoint* – of which, as mentioned in a previous chapter, *MaxIm DL* includes the cut-down version called *LE* (limited edition) that is invaluable in itself. *ACP* includes all three and is essentially a total control

Fig. 8.7 Bob Denny of DC3 Dreams

program. *ACP Planner* would normally be run in advance of any observing session, as will be described.

To install the program from the main Web site – see below – you can select the module of interest, *ACP* in this case (http://www.dc3.com).

The *ACP* link takes you to http://acp.dc3.com/index2.html

Details concerning installation are described on the site. Before *ACP* is installed, *MaxIm DL* must have been installed, configured, and been used normally. *ACP* accesses *MaxIm DL* and commands it to control the camera and filter wheel (where used) within the script. The latest *ASCOM* platform should have been installed. *MaxIm DL* should have been tested with the camera and filter wheel to confirm normal operation.

If you plan to use *ACP's* autofocus feature and have a focuser capable of being controlled by *FocusMax* (free software written by Larry Weber and Steve Brady), this software must be installed and tested as well. Finally, you must ensure that you have at least GSC 1.1 (*Hubble Guide Star Catalog*) installed (350 Mb), or better still, USNO A2.0 (the US Naval Observatory A2.0 star catalog). The latter is over 6 Gb in size due to its total sky coverage down to about magnitude 18 or better. You are likely to want a good star catalog anyway, for use in other programs.

ACP Planner. Before trying *ACP Planner* the author had never spent much time actually preparing the groundwork for an observing session. He would normally use the planetarium software *Guide-8* (or an earlier version) to select targets for imaging during the night, noting their culmination times. Essentially it was always a hit-and-miss session with little real planning. *ACP Planner* changed this attitude to session planning by showing me the possibilities of optimizing telescope time.

The software is actually free and works independently of *ACP* itself, although the resulting *ACP Observing Plans* can literally control a whole session's observing. *Planner* enables you to create a night's observing plan using either *Starry Night* or *TheSky* planetarium software. This method of producing a schedule is a significant leap beyond most automation systems because it allows you to actually visualize your target positions and timing together. It becomes very easy to pick the most suitable targets to image, at what times, and for how long.

The software knows the target positions, though these can be modified should you wish to offset the camera pointing position, such as when wanting to image a group of galaxies using an optimized center (where the center of the image is visually selected for best coverage). It knows when and where to slew, take images, focus, and also guide. An *ACP* observing plan is a formatted text file that consists of a list of targets with image/filter specifications and optional timing information. The plan is used by *ACP* to control the observatory and acquire the requested images.

Plans can be produced using the *Planner* running within *Starry Night* planetarium software. The first target can be scheduled to start at a specific time and a preset exposure sequence already defined. So the finish time is automatically entered. Entry of later targets is similarly accompanied by automatic start and finish entries. Hence, the overall imaging session can be quickly scheduled. Similarly, a set of images using specific filters (red, clear, green, and blue) can be commanded, using appropriate binning settings. The process, once mastered, can easily be repeated for other targets and used on other nights. The final plans can be easily adjusted for use in other sessions if required. The *Planner* includes a myriad of other abilities and features; this short description provides merely a flavor of its capabilities.

To prepare for an evening's session, you make a list of the targets that you want to image; this is perhaps one list for asteroids and comets and a separate list for long-exposure imaging. There is no reason why any number of lists should not be combined, but it might be best to do one type of imaging at a time. Many astronomers routinely prepare the entire night's observing run in the form of a plan, but if your observatory dome has to be manually rotated you might prefer to do shorter runs.

Astrometry is extremely efficiently dealt with by automation. One plan can include perhaps ten asteroids, allowing 1 min per exposure and 2 min for slewing; this conveniently takes about 20–30 min, by which time a repeat run can then be initiated. Within 90 min you can have ten full sets of data (three measures per target) to analyze. Similarly, supernova searching runs use prepared scripts that are filed for repeated use and simply called up when required. These search a preferred region of sky and collect perhaps 10–15 images of regions already stored in your collection of selected images – your database. Comparison of the old and the new is then fairly straightforward; see the later summary of *PinPoint*.

ACP observatory control software. You can prepare a plan for an observing run without needing to know how to write software, or indeed without having either

Starry Night or *TheSky*. Your plan can provide a list of targets for the night, for a specified period, or for immediate use. It can control the telescope, filter wheels, set up exposures on the main and guide camera, position every frame at the center of the target, control your dome shutter and rotation (if used), learn how to improve pointing accuracy, interpret a weather monitor, and close an automatic observatory should the need arise. (You have to make your own coffee!) *ACP* suite does far more than this, which is why we shall cover only basic ways to use it and indicate where those with more demanding applications can achieve their goals.

The first time that the program (*ACP*) is run, you need to enter known equipment parameters in the *ACP, Preferences* tab. Although this could appear daunting, several of the tab entries are not required at this stage. You *do* need to enter basic data about your observatory location and name (for inclusion in any discovery reports), the telescope's optical configuration, guider telescope details and preferred settings, and file locations such as the folder where your star charts are kept, such as USNO A2.0. There is an important point to realize here: *ACP* is going to control your telescope via *MaxIm DL*, so all the proposed operations must be fully capable of being actually performed by *MaxIm DL*. As an example, you can have *ACP* initiate a complete guiding session, but to do so you must have already had *MaxIm DL* routinely guiding on selected guide stars.

ACP will perform the task of selecting a guide star even better than *MaxIm DL* alone because *ACP* will do a precise analysis of the guide star field and try to ensure that a valid star – not a hot pixel – is chosen. Taking this further, *ACP's* actual guiding will start with an initial exposure that you can preset optimally and will gradually increase this exposure to your preset maximum – if a suitable guide star is not found. If this process fails, *ACP* can proceed to image regardless unless you select otherwise. This process illustrates the greater flexibility of your automated imaging session.

ACP documentation. All the components of ACP are very heavily documented, and it is essential to read the various introductory sections carefully in order to appreciate both the huge capabilities of the program as well as the recommended starting procedure. There are video tutorials available as well as the built-in help feature.

ACP in operation. For the very first run there is an extensive do-this-first list of 12 steps in the *Getting Started* section. This sequence gives you the best possible chance of getting it all working smoothly, starting with basic data entry, through verifying connections and getting a test image all the way to enabling the pointing corrector. *ACP* has some prepared plans for completing preliminary tasks once you have defined the required telescope parameters. Much of *ACP's* logic is written in various Javascript and VBScript programs (scripts) that you can run using *ACP's* controls. The scripts include *CalibrateGuider, FindBrightStar, SyncScope, TestPointing*, and, of course, *AcquireImages* (the script that reads observing plans and acquires the requested images), among others. The titles of these scripts are self-explanatory.

Before doing any *ACP* run always ensure that *SyncScope* has been run successfully, or that at least one sample image has been successfully solved within the program.

In a typical scripted run, *ACP* will drive the telescope to the first target, take a sample image exposure (Select an exposure that reaches down to about magnitude 16) and solve it to identify or confirm the telescope's exact position on the sky. If this position is not accurate to within a preset limit (perhaps about 2 arcmin) *ACP* commands a small correction movement. It then activates the guider by taking a sample guide image and actually analyzing it to identify the most suitable guide star. This is clearly working at a far more detailed level than *MaxIm DL* does on its own. It then passes the selected star information to *MaxIm DL*, which then starts the guiding.

When guiding errors have stabilized and reduced to a preset level, the main camera exposure is started. In the case of asteroids, do not generally activate the guider because the exposures are relatively short, so the process of exposing for (say) 60 s happens immediately following any position correction of the scope.

The author's 10-min main camera exposures are usually completed as a sequence of perhaps 6 (1-h worth) in a row, but on other occasions the next target will be selected from the script and an automatic *goto* issued. The only intervention required of the user is to ensure that the observatory dome is moved as required. Those using wheel filters can have the additional changes performed automatically. In fact, as discussed in this section, features too numerous to mention, let alone discuss, are available within *ACP*.

List 8.2 was produced in a minute or two using the *Planner*. The author identified the next two galaxies for imaging, selected the exposure time (interval 600), and set the program to produce three of these 10-min images. Such plans can be retained for simple editing if required. It is also easy to specify the actual time that *ACP* initiates the exposure.

Astrometry in ACP

Frankly, astrometry – the measurement of the accurate position of asteroids and comets in relation to the background of stars – is a real pleasure to do with *ACP*. You can begin by identifying all the asteroids and comets previously indicated as suitable using the *Guide-8* planetarium program. These asteroids are often grouped together. I then obtain their orbital elements from the Minor Planet Center's Web site for asteroid hunters (see below for URL). At the MPC you can request a set of asteroid elements and usually retrieve these within a few seconds and save as a text file – see List 8.3. It is then easy to prepare a plan for *ACP* – see List 8.4 – that involves taking a set of three images of each asteroid or comet such that the first run collects images of 60 s each of several asteroids, and then repeats this entire process twice after a period of perhaps 30 min.

Minor Planet Center Ephemeris Service Web site: http://www.cfa.harvard.edu/iau/MPEph/MPEph.html

This entire process is, of course, totally adjustable, and you can add a 10-min deep sky target between a sequence of asteroid/comet images. It is entirely possible to have a succession of asteroids such that at the end of the first run, enough time has elapsed to enable the second run to be started. Significantly, *ACP* calculates the current position of each asteroid just prior to commanding the telescope movement, thereby ensuring that the latest position is obtained. Before getting *ACP* the author had to identify the position of an asteroid or comet using *Guide-8* and then try to

List 8.2 Simple text plan taking two sets of three images of two galaxies. This plan was generated by ACP Planner

```
; For: Lawrence
; At: 06-Jan-2007 19:47:18
; Location: observatory
; Coords: Lat=50° 54′ 22″ Lon=–01° 23′ 25″
; Targets: 3
; Start: 06-Jan-2007 19:50:03 (local)
; Imaging: 01:45:11
;
; --------------------------------------------
; === Next is M81 ===
; Entire image set for M81 will be repeated three times
;
#interval 600
#binning 1
#repeat 3
M 81
;
; === Next is NGC3077 ===
; Entire image set for NGC3077 will be repeated three times
;
#interval 600
#binning 1
#repeat 3
NGC 3077
;
```

List 8.3 Typical cometary elements obtained from the MPC

```
0085P 2008 12 16.3648 1.147474 0.775283 53.5814 343.4512 4.2171 20081130
  0.0 0.0 85P/Boethin
0144P 2009 01 26.8601 1.439021 0.627787 216.0966 245.5605 4.1092 20090109
  0.0 0.0 144P/Kushida
```

List 8.4 Converting elements into an ACP plan

```
; comet at 6pm
#interval 60
#repeat 2
#binning 2
0085P 2008 12 16.3648 1.147474 0.775283 53.5814 343.4512 4.2171
  20081130 0.0 0.0 85P/Boethin
0144P 2009 01 26.8601 1.439021 0.627787 216.0966 245.5605 4.1092
  20090109 0.0 0.0 144P/Kushida
```

match the field of the sample 10–20 s image to the known position. Needless to say, this added many minutes to the process of obtaining even one measurement.

List 8.4 shows the small additions to the basic element sets required to complete the *ACP* plan. The semicolon (;) informs *ACP* that the line is a comment, the hash (# command *interval*) tells *ACP* that the required exposure is 60 s. The next hash (# command *repeat*) instructs *ACP* to repeat the exposure twice – not strictly necessary, but it sometimes improves the chances of success! The final hash (# command *binning*) is also self-explanatory. With a long list of asteroids and comets I do not normally repeat the exposure as was done here.

An extremely large number of variables is possible for these plans. A change of exposure time (the interval parameter) can be inserted before any target to change that default. Filters can of course be specified; in fact, so many parameters can be specified that it can be difficult to select one rather than another. Check out the DC3 forums for users' comments and questions, and consider a trial period if the advanced nature of this software seems appropriate for your applications.

Testing the plan. Before running any plan, test it. This is a 5-s process! Included in the *ACP* suite is *ACP Plan Checker*. Merely drag and drop the plan (the text file) onto the program's icon, and it will be analyzed to ensure that it is valid. Any errors are immediately reported and are normally extremely easy to fix. A common error is the absence in the database of some faint, remote galaxy or other deep space target, necessitating checking with *Guide-8* to identify the appropriate RA and declination to enter into the plan.

Visual Pinpoint

This program is part of the *ACP* suite and has at least two specific applications. *MaxIm DL* includes the *Pinpoint LE* module that can solve images, display the location of the true center, the measured focal length, and camera orientation relative to the right ascension axis. Within *ACP* this can be used to do a check after a slew to ensure that the telescope is pointing exactly where specified, or within a predetermined error of it. *ACP* can then reslew, if required, to the specified target, ensuring that an accurate centering has been achieved. *Visual Pinpoint* can compare images of the same region of sky taken earlier in the session (for asteroid detection) or at an earlier date (for supernova detection), and "blink" them – at a selectable rate, of course! This blink process can show any changes in position of any objects within the common area of the images. It is therefore an ideal asteroid and supernova detection program. It can also do much more than this.

You set things up by entering all the known parameters of your system into the preferences section in the *ACP* module. There are multiple tabs available, of which some must be edited at the start. Some settings can remain at their defaults; enter the reference path of your catalog for your computer and select the appropriate star catalog. Enter the maximum magnitude reached by your telescope/camera system for these exposures. The Plate Parameters must be known from previous measurements, such as plate solving via *PinPoint LE*. This is clearly essential data – see Fig. 8.8 – that enables an image (plate) to be solved.

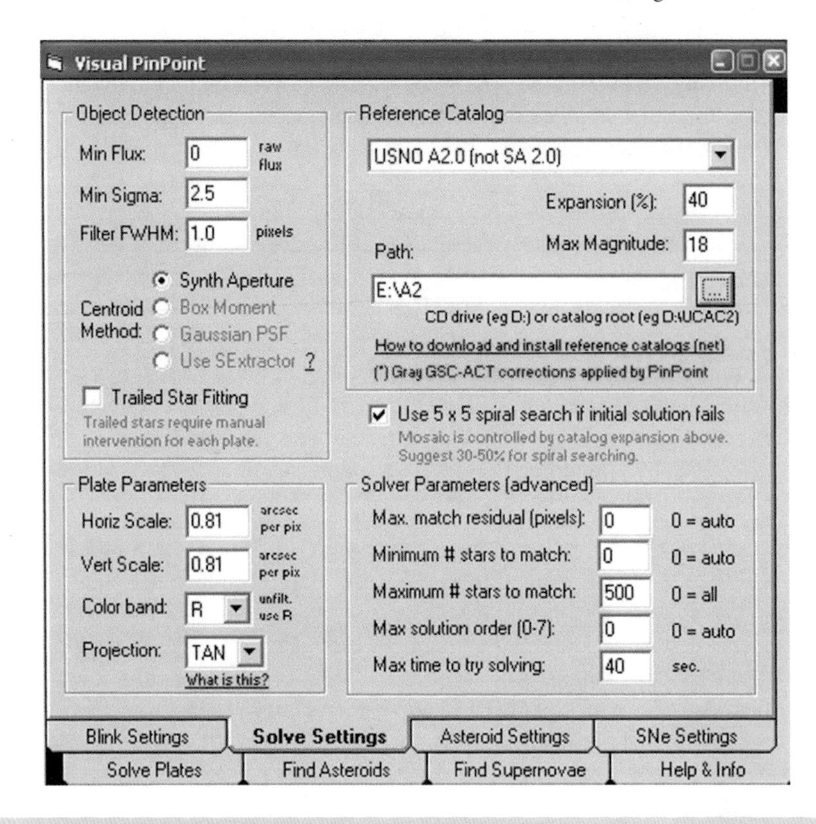

Fig. 8.8 *Visual PinPoint* solve settings tab for pixel and other parameters

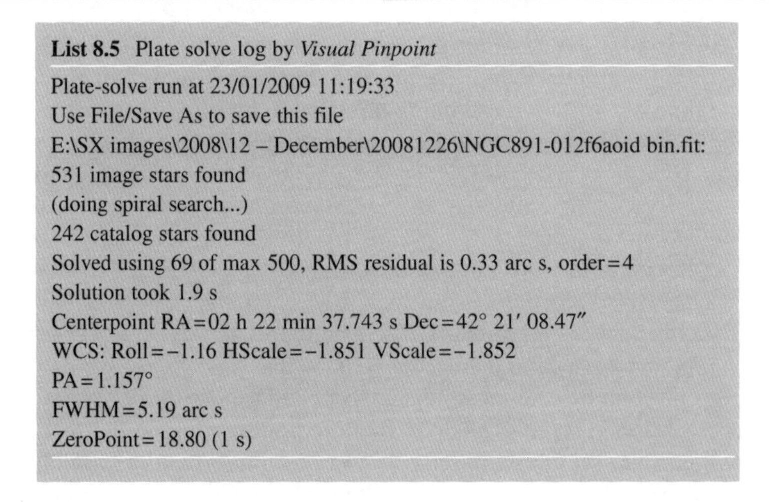

List 8.5 Plate solve log by *Visual Pinpoint*

Plate-solve run at 23/01/2009 11:19:33
Use File/Save As to save this file
E:\SX images\2008\12 – December\20081226\NGC891-012f6aoid bin.fit:
531 image stars found
(doing spiral search...)
242 catalog stars found
Solved using 69 of max 500, RMS residual is 0.33 arc s, order=4
Solution took 1.9 s
Centerpoint RA=02 h 22 min 37.743 s Dec=42° 21′ 08.47″
WCS: Roll=−1.16 HScale=−1.851 VScale=−1.852
PA=1.157°
FWHM=5.19 arc s
ZeroPoint=18.80 (1 s)

List 8.5 shows the result obtained following a plate-solving event. The log shows that 69 of the 531 detected stars were used to provide the resulting data. This image was originally a 10-min full resolution image to be used for combining with others

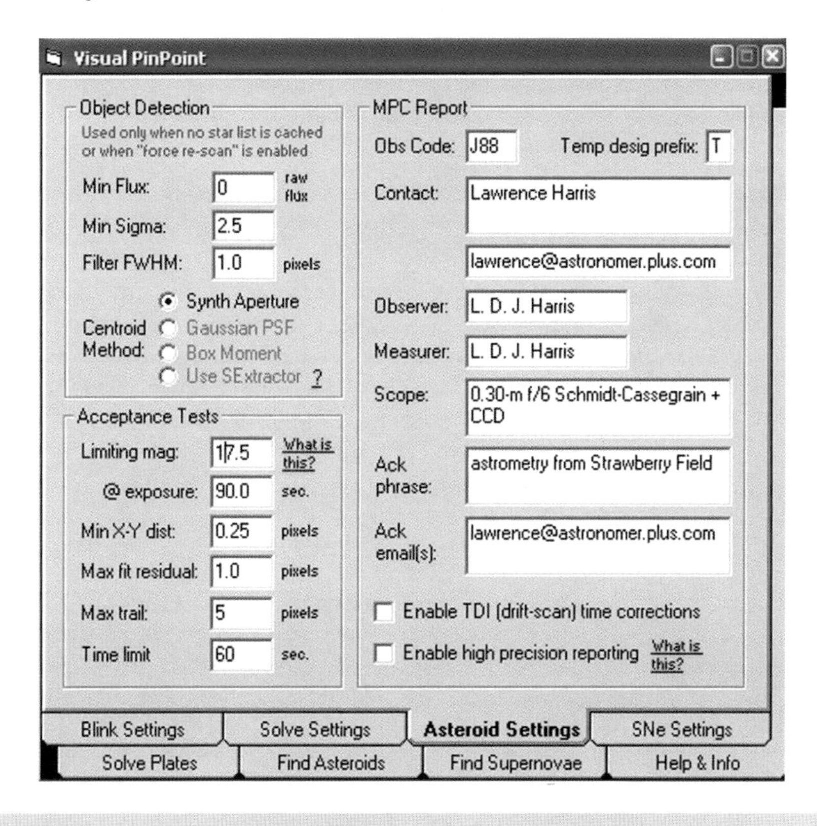

Fig. 8.9 *Visual PinPoint* asteroid settings tab

to produce a long exposure result (2 or 3 h of data). The process is illustrated here by the analysis of one binned image. These results enable you to further tune your settings; the extensive help file explains in clear terms how to optimize the detection and other settings for your own system.

The *asteroid setting* parameters tab should also be completed (see Fig. 8.9). Part of this includes your (approximate) limiting magnitude. Similarly the supernova (SNe) settings tab should be checked before starting your supernova runs.

Multitasking!

After an asteroid or a supernova run it is a common practice to have the telescope start another plan (or, of course, carry on with a longer planned program built into the asteroid run) while you analyze your incoming astrometry. Early analysis ensures that you can quickly identify any new objects during the session. This has been important for devoted supernova hunters on numerous occasions. Such discoveries have to be urgently transmitted to the Minor Planet Center (in the case of a suspect asteroid discovery) or to the International Astronomical Union (for suspected supernovae). In at least one case, the first report of a suspect supernova was only a few minutes ahead of the second report from another supernova hunter!

Summary

The *ACP* software collection includes everything that the keen astronomer could want to use. As has been discussed, it provides a set of facilities enabling one to effectively plan a night's observing, as well as including comprehensive monitoring and analysis tools. The software is available on a 60-day trial that should serve to offer you every opportunity to see whether it can work for you.

CCDCommander

This program was written by Matt Thomas (see Fig. 8.10) and is another of the few programs that control both telescope and camera. Like broadly comparable programs it does require other software to directly control both systems, and then issues the appropriate commands to each. *CCDCommander* can produce images from multiple targets, can dither when requested, analyze a guide camera image to

Fig. 8.10 Matt Thomas

select a suitable guide star, focus automatically (assuming that your telescope system has a suitable focus control facility), and ensure that your targets are centered.

Following installation, the control devices are set using the *Options & Settings* menu and selecting the relevant tab. Other equipment and software specifics are then selected on the appropriate tabs. There is a full help file feature that carefully takes you through the necessary initial procedures. The *AutoGuideStar* tab offers an excellent choice of settings, including *ignore 1-pixel stars* – designed to exclude hot pixel selections.

If you have an automated observatory, *CCDCommander* can control that as well. With a setup plan for a night's imaging, the program can, in principle, take suitable flat images, collect darks, and spend the night controlling the telescope and camera, moving steadily from target to target until early morning light arrives. Some people are prepared to leave the whole observatory under computer control throughout the night, but if your observatory dome is manually operated you cannot do this. Even so, it is still good to be able to set up a session lasting perhaps two hours and be confident that two hours' worth of images will result. At that point the next session can be started after moving the dome around.

CCDCommander currently works with a selection of advanced software, some of which is specifically required such as *TheSky (Professional edition)* or *ASCOM Telescope Controller* (recent versions), *CCDSoft (recent editions), or Maxlm DL (v4.11 or later)*. Other software listed on the Web site can control a suitable motorized focuser under *FocusMax* and others. An instrument rotator, cloud sensor, and dome controller are available. Check that your own accessories can be controlled this way.

Download and trial. CCDCommander can be downloaded from the Web site shown below. There is a trial period of 45 days that should enable anyone to configure the software to match their telescope/camera combination and have several successful sessions. At $99 the software appears to be good value, even for those without an automated observatory or focuser. Ensure that you have the required essential software before installing http://ccdcommander.com/index.html.

MPO Connections

Straight from the Minor Planet Center comes *Minor Planet Observer Connections*, a dual control program for a telescope and CCD camera. Figure 8.11 shows the main screen. Installation of the software produces a large help file in PDF format, and this guides you through the configuration process. At the time of writing, the program does not currently cater for *Starlight Xpress* cameras, but this is not ruled out for the future. Some of the SBIG models and other cameras are included. Several telescope types are available for control, including the LX200GPS and LX400 models. Filter wheels and focus control are available as well as a number of

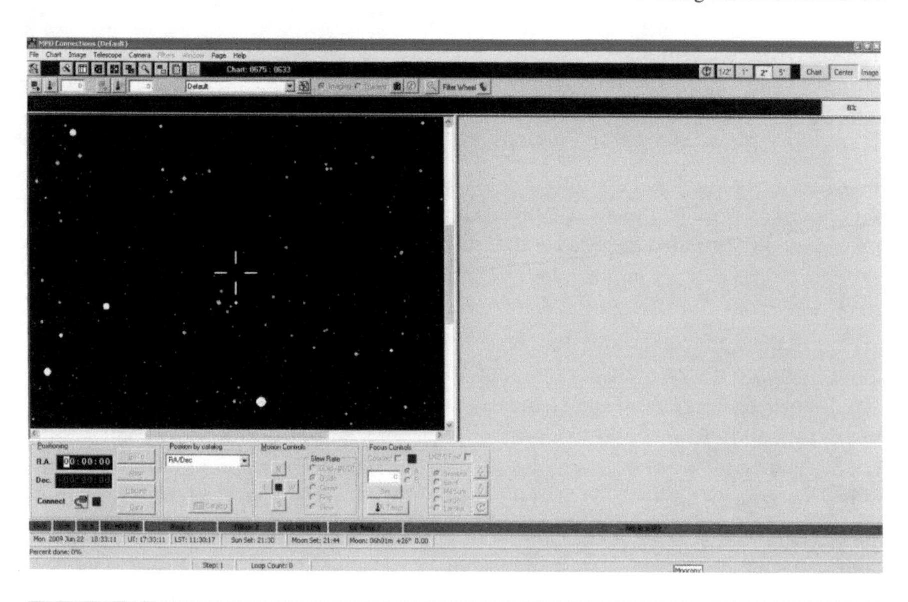

Fig. 8.11 *MPO connections* main screen

features such as multitarget imaging. The user manual can be downloaded in advance so that you can study the features on offer (http://www.minorplanetobserver.com/MPOSoftware/MPOSoftware.htm).

CCDAutoPilot4

This is another program from the *CCDWare* stable. Version 4.0 includes the tools and capabilities required to optimize the imaging system and fully automate image acquisition over the course of an evening. The software includes a number of tools such as camera noise parameter measurement, guider parameters, several system tests, support for *ACP-Planner* (see earlier), mosaic targeting, and a large number of desirable features for astronomical image collection sessions. It would take a considerable time to review the whole of this highly automated program.

As with comparable programs it provides camera control via either *CCDSoft* or *MaxImDL*, telescope control via *TheSky6* or *ASCOM*, focus control via *CCDSoft @ Focus2* or *FocusMax*, and also camera rotator support for the *Optec Pyxis, RCOS PIR, Astrodon TAKometer, or ASCOM*. There is also dome control support for a number of programs. It is $95 for the basic edition and $295 for the professional edition. See the Web site for any changes, and for information about the different versions: http://www.ccdware.com/products/ccdap4/

Fig. 8.12 Craig Stark, writer of *PHD guiding*

PHD Guiding

Craig Stark – see Fig. 8.12 – wrote this software and, according to its Web site (see address below) the title stands for "Push here dummy!" This is basically because all the necessary calibrations are taken care of automatically. You do not need to enter details about the orientation of your camera, nor the image scale. The automatic calibration routine takes care of this; in fact, you may never need to enter any parameters.

The program uses a very simple main display (see Fig. 8.13). The help file is good and provides comprehensive notes. On a first run, you should be able to get the program to perform a guide star calibration and also to guide on a star very easily (see Fig. 8.14). Because guiding is done via *ASCOM*, the usual range of telescopes can be used, as well as cameras. Do be aware that only the guide camera is used. This software is free, although the associated Nebulosity and other software mentioned is not (http://www.stark-labs.com/phdguiding.html).

Additional Software

GPS Control2

This software sets up a virtual hand controller on your computer and supports most of the *AutostarII* functions. The Web site is http://www.interactiveastronomy.com/

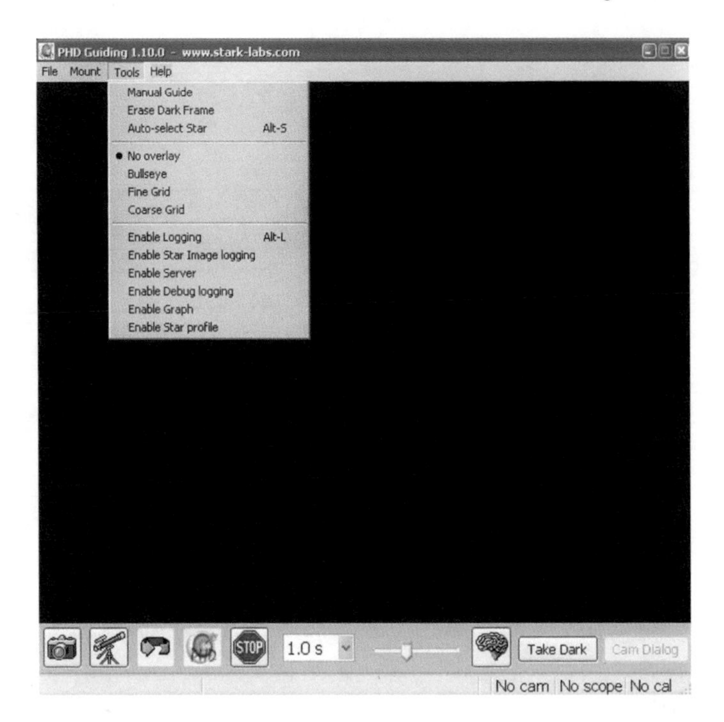

Fig. 8.13 Main screen of *PHD guiding*

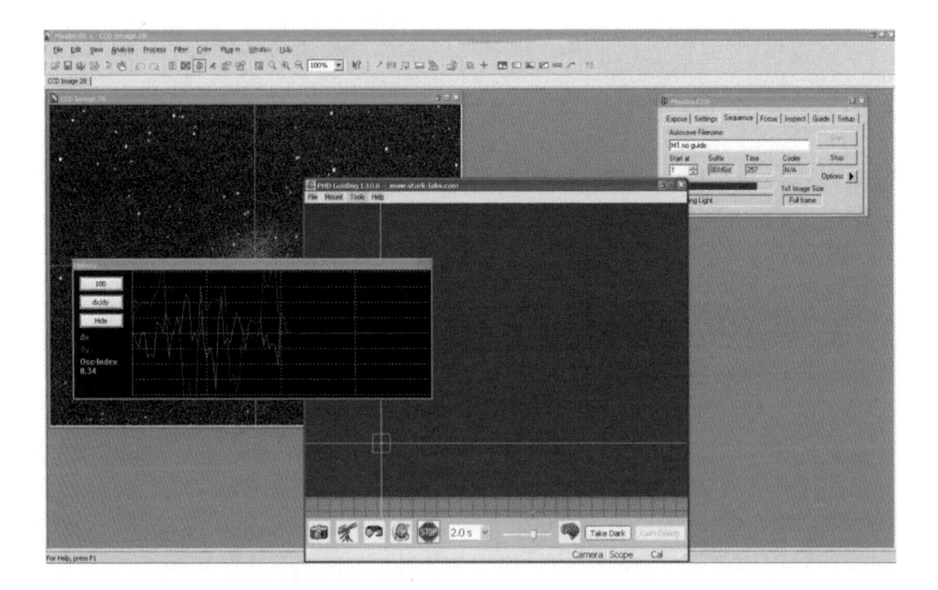

Fig. 8.14 Guiding on a star

SkyTools2

This software is more of a suite that includes considerable planning features as well as being *ASCOM* compatible and therefore capable of controlling your telescope. The Web site is http://skyhound.com/realtime.html

TheSky (v6)

TheSky6 is another comprehensive suite that includes extensive planning features, a detailed planetarium component, and full telescope control facilities. Unusually, *TheSky6* includes direct telescope drivers, including both the LX200GPS and LX400 series scopes. The website is http://www.bisque.com/help/v6/TheSky_Version_6.htm

CCD-Navigator

This is a planning program that helps the astronomer schedule an evening's imaging session by exploring more than 23,000 deep sky objects from 18 celestial catalogs. There are over 500 full-color thumbnails to help you choose your targets. The Web site is http://www.ccdware.com/products/

CCD-Stack

This highly competent image processing program processes multiple images to remove undesirable artifacts such as cosmic rays or airplane trails from the final composite image. The program implements several data rejection procedures to detect unwanted artifacts on a pixel by pixel basis. The program includes sophisticated star bloom removal, image registration, normalization (using weighted values), advanced gradient removal, and deconvolution to improve image sharpness.

Chapter 9

Adaptive Optics

Previous chapters have explained in some detail how all telescopes, including the Meade LX200GPS and LX400 series, require careful setting up and alignment if you are to get the best from them. Before considering further hardware, do ensure that you have taken your telescope to its peak performance by optimizing its periodic error (PEC table), reducing the backlash to a minimum level, and polar aligning your scope to a good accuracy.

Even after such effort, the inherent nature of the rapid gear errors ensures that the telescope mounts cannot cope with high-frequency changes any more than can professional telescopes. Top-end mounts often have considerably higher specifications than their mass-produced competitors, but even these costlier mounts have some level of imperfection in their mechanical components. Like many popular scopes, therefore, the LX series cannot reliably respond to extremely short timescale movement commands because they have an inherent settling time following each individual command. None of these limitations prevent amateurs or professionals from obtaining very high-quality images from the telescopes; they merely establish an upper limit when used with conventional hardware.

Limitations of Guide Scope Use

Despite the limitations, many amateurs produce excellent long-exposure images by using various established methods. One method involves stacking a number of relatively short-exposure images to produce a near-equivalent long-exposure image. Low-noise cameras make it possible to stack exposures even as short as 30 s, although after optimizing, an LX series scope with camera should be able to produce perhaps 2-min multiple exposures without any form of guiding. By using properly set up autoguiding with a rigid guide scope mounted on the main scope,

L. Harris, *So You Want a Meade LX Telescope!*, Patrick Moore's Practical Astronomy Series,
DOI 10.1007/978-1-4419-1775-1_9,

multiple individual exposures of 10 min should be routinely easy to take with few failures, as described in previous chapters.

It is better to avoid attempting to correct for atmospheric seeing – a process otherwise known as *chasing the seeing* (see Chap. 7). Therefore, if you want to take the possibly ultimate step in the quest for a better image, you must aim to try to correct reliably for short-term seeing changes. Active and adaptive optics could be the answer.

Active and Adaptive Optics: The Principles

The terms *active optics* and *adaptive optics* began to appear in astronomical literature several decades ago (around the middle of the last century). Active optics is a new technology designed to simulate reflecting telescopes of very large diameter by combining the optical powers of several smaller mirrors effectively into a much larger mirror. By fitting each component mirror with an actuating device to actively control the positioning of the mirror, a multiple-mirror telescope can behave effectively as a single optical system. Such well-known instruments as the New Technology and Keck telescopes now incorporate this method of control via active optics. Before active optics, the diameter of a large telescope mirror was limited by the sheer weight of glass required to ensure that the scope would keep its shape during slews across the sky. The Hale telescope at 5 m diameter was effectively the largest possible size that could be controlled as one mirror.

The new generation of active optics telescopes uses several small, thin mirrors that are kept rigidly in the correct shape by an array of actuators behind the mirror. This segmentation prevents almost all of the gravitational distortion that occurs in the conventional large, heavy mirrors. A computer program moves the actuators in real time, ensuring that the best possible image is obtained – hence the use of the term active optics. Such systems respond to changes of balance as the telescope moves across declinations and right ascensions, and can also allow for wind buffeting and even mechanical deformations in the telescope structure. The actuators can correct all parameters that may affect image quality within timescales of about 1 s or more, maintaining a nominally optimal shape.

In addition to the causes of degraded image quality discussed in previous chapters, the atmosphere too imparts a highly significant effect on the image reaching the camera. The adaptive optics idea was designed to operate on a much shorter timescale in order to compensate for these effects. They affect the image at much faster timescales (down to timescales of 0.01 s or less), and such variations are not easily corrected with primary mirrors. The new technology was developed for use with small corrective mirrors and secondary mirrors.

By measuring atmospheric distortion using an adjacent detector and rapidly compensating for it, adaptive optics can significantly reduce this distortion. It was only advances in computer technology during the 1990s that made the technique possible, even though the theoretical concept had been understood back in the 1950s. In one form of adaptive optics, a rapidly moving tip-tilt mirror makes small rotations around two of its axes, effectively correcting the positional offsets for the image. Much of the atmospheric aberration can be removed in this way. Such controlled mirrors can therefore improve image quality that would otherwise be significantly impaired because of atmospheric seeing.

One Guide Camera

It is important to note that at any given time, there is only one guide camera in the autoguiding system. The usual guide camera mounted at the end of a guide telescope is not used in an adaptive (or active) optics system. Guide signals come from just one camera. If you look carefully at Fig. 3.1 (Chap. 3), you may notice that when this picture was taken, the guide camera attached to the guide scope was connected to the main camera. The guide camera attached to the adaptive optics unit was not connected.

Professional Application

At Palomar Observatory, the Hale telescope now has an adaptive optics system. This uses a star as a calibration source and then deforms a small mirror to correct for distortions caused by the atmosphere. These corrections are faster than the atmosphere itself can change – some thousands of times per second. The technique has been used for a variety of observations, including long-term observations of the atmospheric changes of the outer planets, studies of the weather on Saturn's moon Titan, the hunt for brown dwarfs, and the details of star formation.

In many instances, there is no suitable guide star to measure atmospheric turbulence; however, by projecting a narrow sodium laser beam into the sky (see Fig. 9.1), astronomers create an artificial laser guide star for use in adaptive optics whenever required. At an altitude of about 60 miles, the laser beam makes a small amount of sodium gas glow. This serves as the artificial guide star for the adaptive optics system. The laser beam itself is too faint to be seen except by observers very close to the telescope, and the guide star it creates is even fainter. It cannot be seen with the unaided eye, yet it is bright enough to allow astronomers to make adaptive optics corrections. Figure 9.2 shows the huge improvement from an active optics system coupled to the Palomar telescope.

Amateur Equivalents

It is evident that the ability to apply such high frequency changes that enable us to follow atmospheric scintillation must result in the best quality images obtainable by amateur astronomers. Of course, in amateur telescope operations, the equivalent technology has to be of a simpler, and less costly, nature. At least two telescope

Fig. 9.1 Hale telescope showing a laser beam penetrating the night sky. (Image courtesy Palomar Observatory)

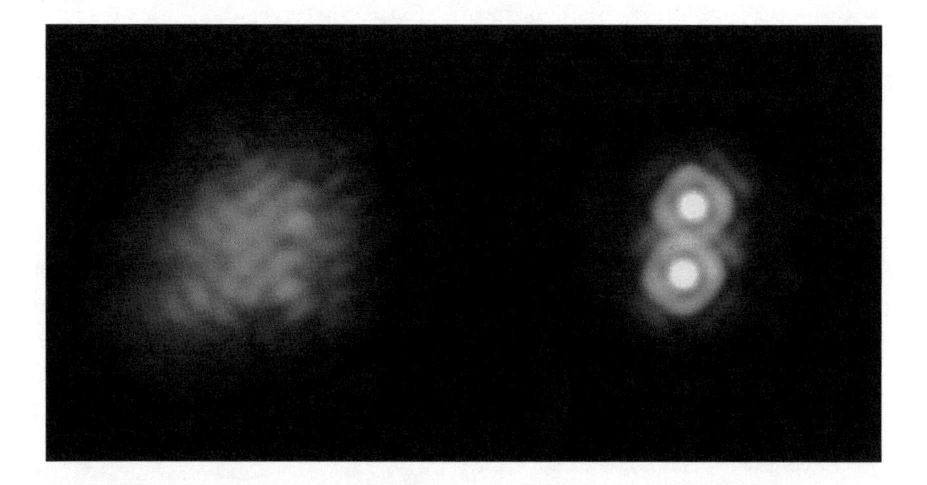

Fig. 9.2 The binary star IW Tau is revealed through adaptive optics. The stars have a 0.3 arcs of separation. The images were taken by Chas Beichman and Angelle Tanner of JPL. (Courtesy Palomar Observatory)

accessory companies have developed their own equivalent of the adaptive optics principle – Santa-Barbara Instrument Group (SBIG) in the United States and Starlight Xpress (SX) in the UK, both units being available in both countries.

Fig. 9.3 *Starlight Xpress Adaptive Optics* unit fitted to telescope. Additional cables are used to control the tilt plate

Santa Barbara Instrument Group: http://www.sbig.com/
Starlight Xpress: http://www.starlight-xpress.co.uk/

The author bought the *Starlight Xpress Adaptive Optics* unit – see Fig. 9.3 – in 2005, and after initial problems caused by the misinterpretation of cable connections and other uncertainties, managed to get it working somewhat later. Although there are some extra considerations that have to be recognized – and which are discussed in this chapter – the unit routinely provides the highest quality images that one could have anticipated. It is difficult to imagine not using it for long exposure imaging.

Starlight Xpress "Active Optics" Unit (SX AO)

This unit should perhaps be referred to as an adaptive optics system, so we will use this term throughout this chapter. In fairness, one respected expert suggested that the term adaptive optics should be restricted to devices providing corrections at higher speeds; the term *tip-tilt autoguider* was suggested. We hope that the reader will, however, accept this use of the term adaptive optics.

One method of rapidly adjusting the position of an image is by using a tip-tilt mirror to reflect the incoming light beam through a varying angle (as used in an older SBIG unit). Although this has its advantages, there are also disadvantages, such as the reduction of the back focus distance. Instead of using this method, the

SX AO unit allows the incoming light beam to pass straight through on to the CCD chip, but interposes a refractive glass element (deviator plate) to cause the path changes. This system is optically shorter than that obtained when using the alternative method and has a well-defined optical deviation for a defined input signal, or change of signal.

The AO element is a multicoated, bloomed, plane-parallel optical window, with a thickness of 12 mm and a diameter of 40 mm. The element can be tilted between ±3° via stepper motors, and this diverts the incoming beam by a calculated amount. With an SXV-H9 CCD camera (one of the Starlight Xpress one-shot-color [OSC] or comparable), the maximum movement corresponds to about ±23 pixels. This displacement is independent of the distance between the CCD chip and the AO unit, unlike tip-tilt mirrors.

The unit can complete an image position correction within a few milliseconds, but the real limitation is defined by the brightness of the guide star, and of course the speed at which the image is downloaded by the computer. Control of the AO unit is usually via serial data from an RS232 port. If your observatory laptop does not have this port, use a USB-serial converter. The guide camera is very much a part of the AO unit itself. An off-axis guider (OAG) – see Fig. 9.4 – is fitted to the AO unit such that incoming light passes through the deviator plate, then an optional filter, then on to the main CCD camera. Within the OAG section, there is a cone of light external to the cone reaching the CCD, which is sampled by a small inset prism.

During initial setting up, it is important to ensure that the shadow of this prism does not reach the main CCD. There are therefore a number of adjustable settings available, including adaptor rings to allow rotation of the main CCD camera to obtain the preferred orientation. (Set this close to 0° so that the chip's axis is parallel to the RA axis.)

Fig. 9.4 Off-axis-guider (AOG) assembly showing sampling prism

The construction of the unit provides two recesses where a 48 mm filter can be placed. An example is an IDAS light pollution filter positioned just in front of the actual CCD camera, at the back of the unit. This avoids the guide camera having any extra filtering that might reduce its sensitivity to faint stars. The recess also means that very narrowband filters such as hydrogen alpha can be used to great effect. For those wishing to use an alternative to the Starlight Xpress guide camera, it is possible to use adapters to accommodate other types.

Telescope Focal Ratio

When using a combined (or in this case, a smaller separate) chip, both chips are in the same overall light beam and see the same overall telescope focal ratio. As a consequence, the number of potential guide stars will be reduced at longer focal ratios (such as f8 and above). This does mean that the faster focal ratios (f6 rather than f10) are more likely to produce good guide stars, increasing the probability of a good result during guiding. There is a limit to this process due to the increasing difficulty of finding focus for the guider. Technically speaking, the unit works better at longer focal ratios. Do remember that the whole unit needs to be recalibrated whenever a fundamental change – such as focal ratio – is made. Recalibration takes only a few minutes.

CCD Guide Camera

As will be realized, much depends on the quality and sensitivity of the guide camera. During early testing and optimizing of the SX AO unit, an improved guide camera became available. The newer one has a higher noise level but is binned on the chip and has greater sensitivity. Specifications will only improve as time moves on.

Assembling the Unit

Caution: Avoid dust! After the light beam's exit from the scope itself, every glass surface will reduce the intensity of the final light reaching the CCD camera. The surfaces can be kept reasonably clean if they are not exposed to dust when not in use. When my SX AO unit is removed from the scope, it goes into a plastic container that remains closed and contains small packets of desiccant. Careful examination of the most recent flat image may reveal whether there are enough dust motes to justify cleaning. Reference to the manual will usually provide optimum cleaning recommendations.

The unit can be assembled without difficulty; orientation of the various adapter rings should be completed after the whole is put together. Ensure that the main body of the camera has its long axis horizontal symmetrically placed between the serial ports. The adapter rings can be rotated to permit accurate orientation. Although somewhat time-consuming, once assembled properly, you would rarely need to disassemble except for dust removal from an optical surface. Finally, remember to rebalance the scope. Any significant change in overall balance will affect its performance. If you normally use the unit incorporated in the optical system, it is best to measure and correct the PE.

The exact configuration of the unit, guider, and telescope may also be affected by your choice of software. Software – whether individual programs or one all-inclusive program such as *MaxIm DL* – has to work consistently and cooperatively. Here is a detailed description of what might be a common combination of equipment and software.

Cable Configuration

1. Power to all three systems: telescope, CCD camera, and SX AO unit.
2. USB cables between computer and telescope, computer and camera, computer and guider (SX AO unit). Note that the camera and guider may share the same cable in some systems, as indeed may the telescope and main camera. This is true in both cases for the LX400 system and means that only one USB cable is required to control the telescope, camera, and guider (except under low temperature conditions).
3. Serial cable (via USB-serial converter) from computer to signal input of SX AO unit. This is the cable that the software uses to control the orientation of the refractive glass element within the AO unit.
4. Bump cable from SX AO unit to telescope autoguider input. This cable is used by the software to provide an occasional move (known as a *bump*) in right ascension or declination whenever the tip-tilt plate reaches a preset limit. This bump is used to move the telescope by a small amount in order to bring the deviator plate back within its limits.

Software Connection

The Starlight Xpress AO unit is supplied with its own software. Because *MaxIm DL* is popularly used for observatory operations, we shall describe the process of testing and operations using this application. The supplied software should be found adequate to test your system as well as for some routine imaging.

Make sure that the power and other cables are fitted firmly. You do not want to have cables dragging on the equipment because – as explained in a previous chapter – this can significantly impair guiding quality. Instead of merely leaving the cables loose, fit them, where possible, to handles and other supports that still allow them to freely move with the telescope during both slews and routine guiding.

After all the cables have been securely fixed, rebalance the telescope, particularly the declination axis. In practice, this is rarely needed because you will mostly use the same optical configuration. Allow enough time for your CCD camera's cooling system to achieve stability; generally allow 20–30 min, although less is probably adequate.

MaxIm DL should be started after the USB connections have been made; the necessary drivers are loaded at that time. Open the *telescope control* window and then the *camera control* window. The latter usually takes some seconds to display. Figure 5.3 in Chap. 5 shows the camera being configured – in that instance, without the AO unit being used. The software must be told of the presence of the adaptive optics unit by selecting *setup, setup*, and *AO*. You may well feel uncertain about which COM port you need to enter here, so a good rule is to open the *Windows device drivers* option and then display all COM ports. If you do this when you first connect the USB cables, you can immediately see which COM ports have been chosen for the individual pieces of equipment. You can also do this after starting *MaxIm DL*, though do note that two COM ports may be allocated, and you need to be able to differentiate the equipment.

Hopefully, you should be able to enter the specific COM port number of the AO unit – entry (in my case) COM port 5. A possible problem may arise here. *MaxIm DL* permits the use of COM ports 1–15 for the SX AO device. When I first tried using my new laptop for my observatory, it chose COM port 16! There are two ways of dealing with this, so I learned them both. One way is to see whether a lower order COM port can be used for the SX AO unit; another is to reassign the COM port to a lower, unused COM port.

You should now be able to connect both modules. The telescope module (with the previously selected LX200GPS-R driver) should take just a few seconds; it may produce a notional error about the driver version number. When the camera module is connected, the SX AO unit will be initialized if everything is configured correctly. Initialization involves the tip-tilt window being tested for movement ability; this takes a second or so to activate and test. If connection fails, you may have to carefully identify which part of the CCD camera configuration has not been set correctly. One word of advice: try connecting for a second time (without changing anything). Often, the initialization process only works on the second attempt, so do not abandon hope if the first connection attempt fails!

Focusing the SX AO Guide Camera

When the main telescope has been focused (and only when), the next stage is the individual focusing of the guider. On the SX AO unit, this can be extremely fiddly, though once completed, it should remain operational for a long time – or at least until dust motes reach the level where you decide to dismantle the unit and clean the optics.

The guide camera chamber has two types of adjustments. A nominal length of the barrel has a flat section on one side, so this can slide in and out during early adjustment (while watching the main camera image via the continuous focus feature) until the shadow of the guider has left the area of the main chip. At that point, this adjustment screw can be tightened. Actual focus is then performed by using the guide camera *continuous focus* option and releasing the two tiny grub screws positioned immediately below the barrel that hold the focus position. Carefully watch the focused image of a fairly bright, unsaturated guide star as the barrel focus position is changed. You can use the numerical readout option if you find this more convenient. Eventually, you should find the optimum position. Finally, test the main camera for focus once more, and the guide camera for focus. On occasions, you might find positions where the image is badly distorted but guiding still proved feasible. Aim to obtain clearly focused guide stars.

Dark Frame Calibration of the Autoguider

If you are familiar with *MaxIm DL*'s camera calibration feature, you may already know that every image taken via *MaxIm DL* can be bias, dark, and flat frame corrected. This also applies – at least in principle – to guide images. *MaxIm DL's* manual explains how to prepare and process sets of bias and dark frame images so that, in theory, subsequent light frames taken with the autoguide camera are fully calibrated.

One limitation of using *MaxIm DL* to control the SX AO unit is that (at least with versions 4 and 5) the software does *not* permit dark subtraction of the autoguider frame during actual tracking, even though there is an option to do this! In my early days of configuring the autoguider, many of the images were noisy, though not always excessively. A typical 1 s frame would include a few hot pixels, and *MaxIm DL* repeatedly misidentified these as suitable guide stars. It may also be a surprise to realize that an exposure taken using the guide camera window does not always produce the same result as one taken using the SX AO (guide camera) window! The author discovered this anomaly following many hours of pondering why exposure results were so inconsistent. Some single guide exposures would show full calibration (that is, no hot pixels), yet others showed multiple hot pixels. Eventually, it became clear what was happening and a work around was derived (explained shortly) to ensure that a valid guide star is properly selected for tracking purposes.

Full Calibration of the Adaptive Optics Unit

In order for the unit to calculate the commands required to send to the telescope to provide bump instructions, and to the AO glass element to divert the incoming light by a known amount, the SX AO-telescope combination must be calibrated. There are detailed notes about the process in the online *MaxIm DL* manual, but we are including a description here because of its importance. During the earlier stages of main camera adjustments, you performed a calibration in which a sample bright star was driven across the field, sequentially in RA and then dec, in order to measure the actual movement for known telescope moving commands. Subsequently, this image scale calibration allowed you to point the mouse anywhere within an exposed image and command the telescope to center the image at that point. A further image should show the scope correctly positioned at the selected point. The next stage is a comparable calibration – but of the adaptive optics unit – performed using the guide camera.

Before the actual (SX AO) drive calibrations are done, you need to find a suitably bright calibration star, preferably near the meridian and not too far above the celestial equator. Locate one and position it near the center of the guider window. Activating the *locate* tab should identify the star correctly. It is important to adjust the exposure to avoid saturation; this then allows the centroiding algorithm to correctly identify the brightest pixel in the star. Figure 9.5 shows the level at 255 (saturation), after which the exposure can be reduced prior to doing the calibration.

The *MaxIm DL* display for guide star and general calibration varies significantly between versions 4 and 5. Figure 9.5 shows the SX AO unit performing a version 4 locate and guiding test. Version 5 shows five windows, including: observatory, camera control, and SX AO control unit commands – see Fig. 9.6. The main guide camera controls (seen in the lower right window in Fig. 9.6) perform a guide camera calibration on every normally exposed image (assuming, of course, that a dark and bias calibration was previously set up).

Exposures taken using the *SX AO control, locate* (lower middle window) will normally show *only* the raw image. On rare occasions, it produces a fully calibrated image! Take an exposure using the guide camera window followed by one using the SX AO window to experience this. It can have a significant effect on guide star identification. When fully calibrated (using the main window), any genuine guide

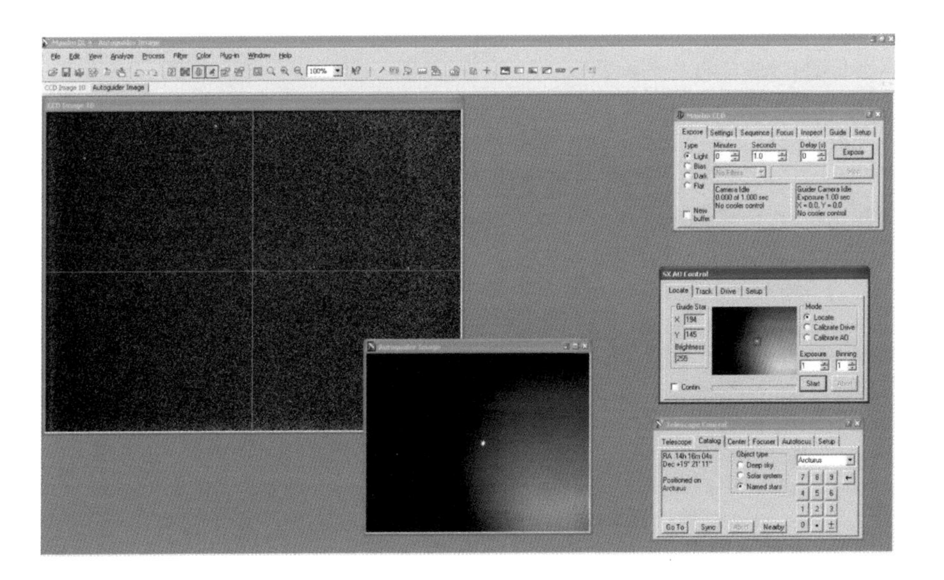

Fig. 9.5 Selecting a bright star for SX AO calibration using *Maxim DL Version 4*

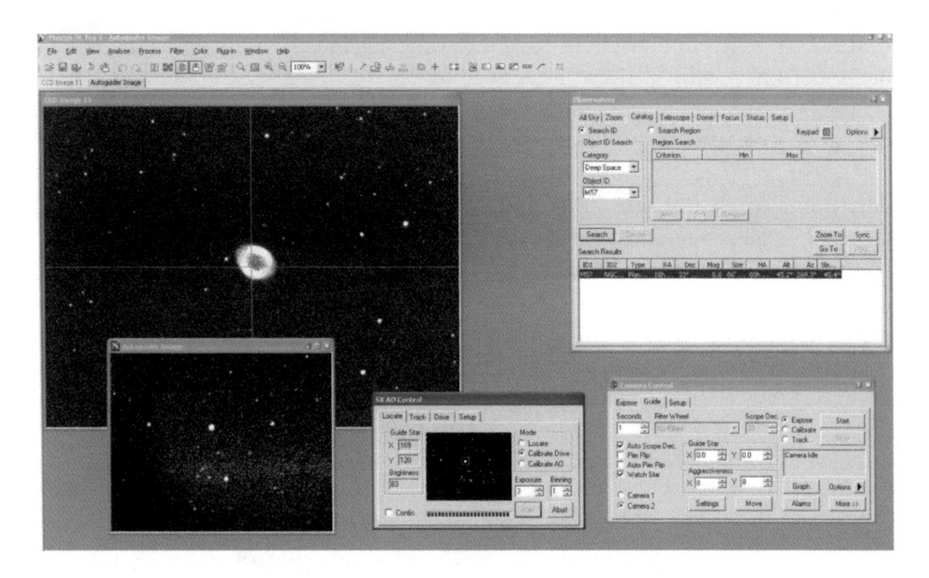

Fig. 9.6 Calibrating the SX AO drive motor using *Maxim DL Version 5*

star may well be identified by the software and therefore have its brightness and position displayed – see Fig. 9.6 showing level 63. Under almost all conditions, an identical exposure taken using the SX AO unit window will be raw and therefore possibly showing several hot – or at least warm – pixels. Unfortunately, these always take precedence over genuine stars and therefore give an unusable result. Consequently, *always* do your initial guide star identification using the *guide camera* window. By noting the location of any suitable guide star, the locate window can then be forced to select the correct guide star.

Bump Calibration

After using *locate* to identify the bright guide star nearly central in the guider window, select *calibrate drive* (*SX AO Control* window in Fig. 9.6). This causes a series of images to be taken while the telescope is driven in Right Ascension and declination. The software then derives the command sequence necessary to bump the mount through small distances along each axis in order to bring the guide star back into the predefined movement limits of the tip-tilt plate. There will always be residual drifts, whether in declination – due to imperfect polar alignment or due to other causes such as changing PE – as discussed in earlier chapters. Any combination of these may eventually take the guide star outside the limits of the tip-tilt window. If by chance the guide star is driven either too far (going off chip) or not enough (less than 5 pixels), then change the calibration time (on the drive tab) for the relevant axis. By careful adjustment of these parameters, a good calibration can be achieved.

Calibrate SX AO

This is a quick process of calibration in which the deviator is exercised in all four directions by the software and "learns" how much to tilt and tip the plate in order to keep up with the changing guide star position.

First Test

When these calibrations are complete, you are ready to test the SX AO autoguiding. Find a suitable target for the main camera, and then find a good guide star for the guider. Guide star identification may or may not be easy. For a first test, the most important factor is to ensure that you have a bright guide star. To run the guiding at a good speed, aim for exposures between 0.1 and 0.5 s. This should really test your system.

Locate: this should automatically select a valid guide star – and it will if you have got a bright star well within the guider field. It is possible to select bin2 on the SX AO unit window should you find that your guide star is not very bright, but it is best to find a bright star for the first test. All being well, your genuine star should be selected and should be bright enough to reduce the guiding exposure to between 0.1 and 0.5 s.

Track: Switch to the *track* tab and set the tracking exposure to between 0.1 and 0.5 s. Select *start* and watch! As long as the calibrations were completed without incident, the guide star should quickly be controlled by the software and kept centered. Monitoring the RMS (tracking error) indicators should be rewarding when you see the initial figures reduce and stabilize at about 0.1×0.1, indicating that guiding is within 0.1 pixel for both axes!

Expose: When reliable autoguiding is underway and working well, you can start the exposure on the main camera. Try a 60-s exposure as a first test, and see the result. You should be delighted! The author's very first test exposure was so much better than those from previous autoguided imaging using the guide scope that it was truly impressive. The stars were much tighter. Note that you can leave the guiding (tracking) running between test exposures. Try a 120-s exposure to confirm that autoguiding continues to perform well. Try a 5-min exposure and then a 10-min exposure, remembering to save each one with a suitable name.

Results and Summary

You will quickly establish your own preferred sequence of operations (as described above) and settings (such as autoguider *aggressiveness* and *maximum pixel move*) and discover the limitations of this method of autoguiding. You cannot always find a suitable guide star within the small field visible to the guide chip. This is probably the first limitation that you may discover. The lower the focal ratio (f6 is better than f10), the more guide stars there should be to occupy the small field. If no suitable guide star is found, here are some steps that can be taken:

1. By binning the guide image and increasing the exposure to about 3 s, you are more likely to find a guide star. Once identified, you may be able to autoguide at between 0.7 and 1 s exposure with the binned image.
2. Familiarity with the extent of the main CCD frame should allow you to drive the telescope, so that although the main target does not remain central, a guide star is far more likely to be found within the guide field. It is possible to derive coverage rectangles for both main camera and guide camera and to have these superimposed on a planetarium star field display to allow possible selection of a nearby bright guide star. On this basis, you may only need to abandon very few targets where no guide star could be found.

Messier 92

One of the results was obtained during a session in which the guide camera had not been fully focused after changing the telescope optics. The track window close-up showed the distorted guide star due to not only poor focusing but also image distortion possibly caused by a noncentralized guide camera! The exposure was started before the problem was noticed. Despite this, well into the 10-min exposure, the RMS wander was showing 0.4 and 0.2 for the X and Y axes, respectively, so it was left alone. Normally, on a steady, clear night, this would be around 0.1×0.1 pixels RMS. However, tracking remained satisfactory, as can be seen from the image of M92 shown in Fig. 9.7. Significantly, the guide exposure was as long as 1 s due to the poor guide star image quality.

Fig. 9.7 M92, a 10-min exposure obtained during above session on June 25, 2008

A close examination of Fig. 9.7 shows that focus is a little soft, but considering that the guide star exposure was relatively long, the stars are still round. This was from a short imaging session around mid-summer. When obtaining a sequence of images, allow at least 5 s between images in order to allow the guide star to be reacquired (assuming that it is not tracked during main image download), and then centered following image download. Also, consider using a dither of 3 or so pixels for later use during image stacking. By slightly changing the position of each frame (the *dither* setting), any hot pixels or cosmic ray hits can be eliminated from the final stack during image processing.

Ready for M27

The next popular target pursued during this particular summer session involved imaging M27. The *goto* M27 feature within *MaxIm DL's telescope* tab brought M27 on to the chip, though not centered. The image was solved using *MaxIm DL's Pinpoint LE*; the scope was then synchronized with the solved position (using *telescope control, sync*) and *goto* M27 clicked again. This movement repositions the nebula at the center of the chip. A further 10-s exposure confirms the image is centered (see Fig. 9.8).

After positioning the main camera on target, luckily there were several potential guide stars available after the *locate* guide exposure. Unfortunately, *MaxIm DL* selected a hot pixel for tracking, so one of the bright stars was manually selected

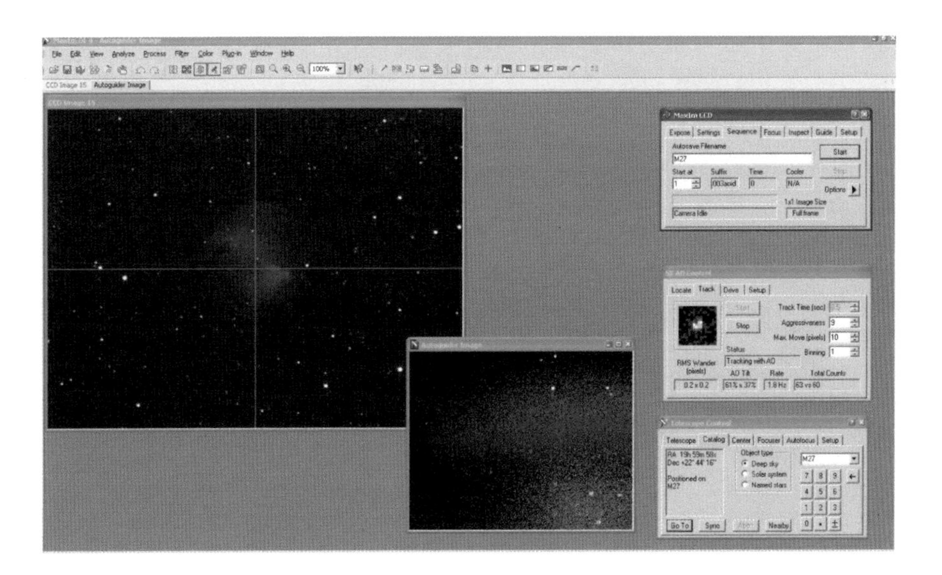

Fig. 9.8 Ready to image M27; 10-s main camera exposure and guide star selected

Fig. 9.9 M27 total exposure of 40 min (four times 10 min) of M27 taken with SX AO unit fitted with IDAS filter

using the workaround described earlier. (Reminder: the workaround involves taking an exposure using the *MaxIm DL CCD guide camera* window because it can invoke dark subtraction, and therefore normally eliminates hot pixels. Then select the guide star, note its position as displayed on the bottom of the screen, and click the *locate* window until the true position is shown.) This manual process takes only a few seconds to complete.

Tracking was started after setting the exposure time to 0.5 s. The software quickly identified the true guide star and pulled it into a stable position, causing the RMS tracking errors to reduce and stabilize. When this was achieved – exactly the state shown in Fig. 9.8 – the main exposure was started. The imaging was then left operating until the completion of the M27 image session. (Note that this session used Version 4.) Four 10-min frames were collected while monitoring the process via a remote network from indoors (see Fig. 9.9). The author's scope can be left tracking for nearly an hour before the dome has to be moved around. As well as routine monitoring, one can do tests, such as changing parameters to see how well the scope reacts.

Background SX AO Settings

The level at which bumping is triggered can be set. During your own testing you can identify your preferred bump settings. Try using 15%, such that when the plate has moved from its neutral position (at 50%) to either 35% or 65% the bump command is issued. You can test different levels of bump duration (for example 10 upwards) and see what setting brings your guide star image back to a nominal 50% or so within a short timescale. The *aggressiveness* parameter controls the percentage of the calculated correction that is actually applied to the plate. You are likely to find that a setting between 5 and 8 will work in most cases. The Starlight Xpress AO manual suggests a method for testing the effectiveness of various settings, or you can watch the tracking process and try changing the settings for yourself.

Conclusions

As is often repeated in this book, it is essential to complete the basic setup of the scope so that the polar axis is accurately aligned, periodic error and backlash in both axes minimized, and the entire scope well balanced. Only then can you test the basic performance of the scope and determine how long it can track accurately enough to obtain good quality images. A guideline figure of about 1 min's unguided exposure without trailing might be the *minimum* to be expected with an LX200GPS or LX400 telescope. As discussed, autoguiding using a conventional Optical Tube Assembly (OTA)-mounted secondary scope is by far the most common method used to extend individual exposures. Atmospheric seeing variations form a natural limit to the majority of amateur-operated scopes, causing virtually all images to have a larger FWHM (full width half maximum) than would be the case if there was no atmosphere. The use of an adaptive optics system enables us to cheat the atmosphere's variability to at least some extent, raising the potential quality of images towards the limit attainable by any amateur scope.

This equipment comes at a cost; just when you believed that you had finally bought all the equipment necessary to achieve near-perfect images, another several hundred dollars becomes "essential!" On the negative side, the devices are relatively costly and no guide star is (currently) guaranteed to be available. In practice, you should expect to find a guide star with 5 out of 6 (or better) targets. This may improve in the future with technical developments, possibly even the availability of artificial guide stars! On the positive side, very few of the author's 10-min exposures are now inexplicably lost, and the quality of the images has never been better.

Chapter 10

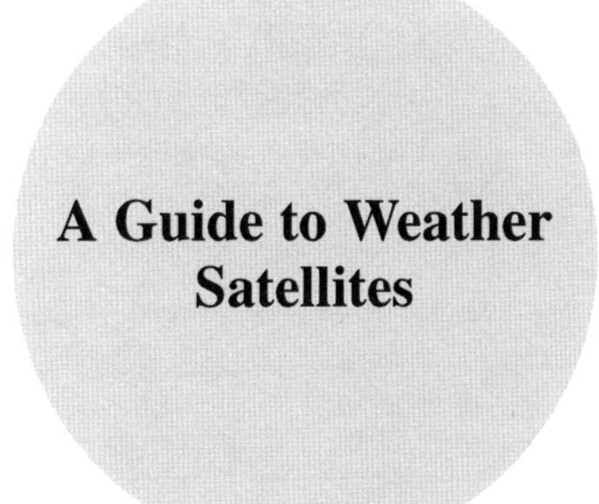

A Guide to Weather Satellites

Whenever there is a live weather satellite image (amateur radio enthusiasts use the term WXSAT) being displayed on my computer, I am reminded of the immortal line from an old comedy by the late Tony Hancock: "There can't be many people around here who know it's not raining in Tokyo!" Although that comment referred to amateur radio hams discussing the weather around the world in the early 1960s, it could now be made by thousands of amateur weather monitors in numerous countries. In this chapter, we take a brief look at the background history of satellite remote imaging, the main constellations of imaging weather satellites, the types of images that they deliver, how they can really help the astronomer, and the options for amateurs wanting to obtain them.

Many people have taken up the hobby of decoding weather satellite signals and producing high-quality images – often out of an interest in the pictures themselves. The images can range from low-resolution image strips generally running from south to north (or vice versa) across your own locality (see Fig. 10.1) to much larger-scale land and ocean masses (see Fig. 10.2), depending on the type of satellite. For the astronomer with an advanced telescope of the LX200GPS type, or a comparable device, the images invite a whole new application. With a perfectly standard weather satellite receiving station, you can routinely monitor the almost live weather situation and can – without any professional knowledge of meteorology – anticipate the near-term weather (known as *nowcasting*). During an observing session, knowledge of the cloud situation for the next few hours is clearly important. Such receiving stations are not for everyone, but you might at least find it useful to know what is available should you wish to consider the possibility in the future.

L. Harris, *So You Want a Meade LX Telescope!*, Patrick Moore's Practical Astronomy Series,
DOI 10.1007/978-1-4419-1775-1_10,

Fig. 10.1 A visible light image from NOAA 17 (see later text about NOAA polar orbiters for a full explanation). UK readers can recognize the clouds that trouble astronomers!

A Brief History of Weather Satellites

Back in the 1940s, rockets had been developed largely for military purposes, but the international scientific community was well aware of their huge potential for space research. The goal of launching the world's first artificial satellite was formally made during the International Geophysical Year (IGY) in the 1950s. The public joke at the time was that this "year" was actually about 18 months long, implying that scientists did not even know how long a year was!

The race was eventually won by Russia with the launch of Sputnik-1 on October 4, 1957, a day that many clearly remember. The Americans subsequently launched a satellite a few months later, under full public scrutiny. This launch was followed by the formation of the National Aeronautics and Space Administration – NASA, a civilian space agency – in 1958.

NASA immediately began planning a number of satellite projects designed to study Earth from the new frontier – near-Earth orbit. From that time in 1958, NASA

Fig. 10.2 A visible light whole disc (LRIT) image from GOES East (12) (coastlines added). See later text

satellites (and those of many other nations) have been studying Earth and its ever-changing environment, using satellites to observe and measure the atmosphere, oceans, land, ice, and snow, and their influence on climate and weather. Initially the emphasis was on satellites equipped to image Earth and help meteorologists forecast the near-term weather. These early weather satellites – the TIROS (Television and Infra-Red Observation Satellites) series – were rapidly improved as more advanced electronics were developed and ever better equipment became available.

The TIROS series spacecraft were just the beginning of a decade-long series of the early polar-orbiting meteorological satellites. TIROS was followed by the TOS (TIROS Operational System) series, and then the ITOS (Improved TOS) series, and later the NOAA series, the current satellites in operation today (from which Fig. 10.1 is a typical example of a section of a received image). The spacecraft were developed by Goddard Space Flight Center and managed by ESSA (Environmental Science Services Administration), this organization having the objective of setting up a global weather satellite system.

From the early 1960s, meteorological, hydrological, and oceanographic data from satellites have had a major impact on environmental analysis, weather forecasting, and atmospheric research in the United States and throughout the world.

NASA research and development saw the instigation of the GOES (Geostationary Operational Environmental Satellite) program within NOAA (National Oceanic and Atmospheric Administration). A series of spin-stabilized geostationary satellites was built and launched, introducing the new satellite service. NASA had two Synchronous Meteorological Satellites (SMS) starting with the launch of SMS-1 in May 1974, and NOAA's GOES series followed with GOES-1 in October 1975.

GOES significantly advanced our ability to observe weather systems by providing visible and infrared imagery of Earth's surface at frequent intervals. GOES data soon became a critical part of National Weather Service (NWS) operations by providing unique information about existing and emerging storm systems both by day and night. Subsequently, more spectral bands were added to the optical scanners, enabling the GOES system to acquire multispectral measurements from which atmospheric temperature and humidity soundings could be derived. Meanwhile, Europe and Asia developed comparable geostationary weather satellites to complement the GOES satellite constellation.

The Cold War Agreement

After 1960, a multinational approach became the obvious means for funding expensive nonmilitary satellite projects. Even in the depths of the Cold War years, Russia and the United States had agreed on a common analog format for the transmission of polar weather satellite images. The transmission format uses a range of frequencies in the 137 MHz band and a standardized modulation method – called automatic picture transmission (a.p.t.) – in which actual image data (dark land and white cloud) would simply amplitude-modulate (a.m.) a frequency-modulated (f.m.) carrier frequency in the 137 MHz band. This unique technique continues today, allowing many thousands of people – particularly amateurs – to decode the image stream using low-cost equipment.

This v.h.f. band image transmission is low-resolution and is derived from an accompanying very high-resolution transmission in the 1,700 MHz band. Although the analog a.p.t. is being phased out – it is being replaced by far more data-intensive digital transmission formats – we can still expect several years of low-cost simple a.p.t. imagery.

Weather Satellite Orbits

The laws of physics (and, in particular, Kepler's laws of planetary motion) ensure that the further from Earth a satellite orbits, the lower is its orbital velocity. Consequently, satellites in low-altitude Earth orbits (approximately 800 km) have short orbital periods of about 90 min. Some of these – the polar orbiting weather satellites – use highly inclined orbits, making several passes each day over the polar regions. At the same time, Earth rotates below, so that the satellites cover the whole planet within 24 h.

The Moon (Earth's natural satellite), at 404,000 km distance, takes 27 days to go once around Earth. In between, as pointed out by Arthur C Clarke in his historic paper on geostationary orbits, there is an altitude (belt) where a satellite orbits once every 23 h 56 min, and therefore remains apparently stationary as seen by someone standing on Earth. This is known as the *Clarke belt*, and the advantages of positioning satellites here – 35,787 km above Earth – were recognized in his paper. In geostationary orbits, satellites remain over the same location on Earth, viewing the same hemisphere 24 h per day. Weather satellites positioned here can therefore monitor the weather continuously.

Because their motion keeps pace with Earth's rotation, you can point your receiving antenna at a geostationary satellite without having to track it – just as you do with satellite television (these satellites are also geostationary). Depending on your longitude, you may be able to receive transmissions from more than one geostationary weather satellite, and many provide virtually continuous transmissions. Over the continental United States, there are two geostationary weather satellites – GOES-East over the east coast and GOES-West over the west coast; both can usually be monitored. Both have an uninterrupted view of the hemisphere over which they are positioned. From any single location on Earth, only those well above the local horizon can be monitored, but with several operating at various positions around the globe, there will always be one for you (see Fig. 10.3).

Fig. 10.3 Geostationary weather satellite positions. (Image courtesy NOAA library)

Being in geostationary orbit does not mean that satellites positioned there achieve all the required goals. For instance, they cannot image the important polar regions. Consequently, a number of carefully selected orbits are used. This is the idea of constellations: a group of similar satellites that do one job in one type of orbit, and a different group of satellites to do another job in another type of orbit. Yet, all have the same overall mission, to observe Earth's weather. For weather satellites, we will be looking at just two types of orbit: polar and geostationary. First, let us look at the types of images that are common to both constellations.

What Do the Images Show?

The weather systems that dominate our lives can be hundreds or even thousands of kilometers across. Although weather radar has extended our ability to see regional precipitation, it is only weather satellites that give us the whole perspective. These satellites – both in geostationary and polar orbiting constellations – carry a considerable number of sensors that are able to measure a variety of weather-related parameters, helping us to build a model from which short- and medium-term predictions can be made about their future development. We can look down on these weather systems and understand the relationship between the storms that we experience and the clear weather that often follows.

Visible Light, Infrared, and Water Vapor Images

The imaging sensors onboard all weather satellites respond to two basic types of radiation. The Sun constantly provides a wide spectrum of radiation, from X-rays to long-wave radiation, but our interest is specifically in visible light and two portions of the infrared radiation band.

Visible Light Images

Visible light satellite images are those seen by reflected sunlight. These therefore look similar to those taken with an ordinary camera, except that they are in monochrome – see Fig. 10.1 – showing a NOAA-17 (polar satellite) image. Clouds appear white, and the ground and sea surfaces are dark gray or black. Naturally, this type of image is only available during daylight hours. During the day, low clouds and fog are normally easily distinguished from land. To some extent, hazy conditions associated with air pollution can also be monitored. During the early morning and late afternoon, shadows of heavy clouds can even be seen on lower clouds or land. Snow cover is easily seen under clear skies because it does not move with the clouds. Some land features, such as mountains, rivers, and cities, can also be identified.

Note that there are no color sensors on weather satellites; everything registers as a shade of gray. These (visible light) images therefore appear as monochrome images. However, color is easily synthesized.

Infrared Images

The infrared spectrum is very wide, so weather satellite instrumentation has been designed to concentrate on specific portions of the spectrum where features can be identified. Earth is always radiating heat. Infrared sensors detect heat energy given off by everything within their field of view. The sensors are carefully designed to respond to temperatures in the range of those found across Earth, whether of land, sea, or cloud. The intensity of this radiant energy depends on the temperature of the emitting surface, so infrared images are really maps showing temperatures across Earth.

Although these temperatures vary during the 24-h period, Earth and atmosphere emit heat both day and night, so infrared images are available continuously – see Fig. 10.4. This means that you can closely follow cloud movements while you use your telescope. Infrared sensors on these weather satellites can distinguish detail as small as 1 km. On these images, both warm land and water can appear dark.

Fig. 10.4 Infrared LRIT image from GOES-E 0531GMT May 7, 2009. (Courtesy Dave Wright of Dartcom)

High clouds are cold and so appear white; lower-level clouds are warmer and so appear gray. Low cloud and fog often have comparable temperatures, so they are frequently difficult to detect in the infrared when their temperatures are nearly the same as nearby Earth surfaces.

It is a common practice to enhance infrared imagery, even by amateurs. Images can be processed to emphasize temperature details by defining contrasting grays and color to match suitable temperature ranges. The results are often used on television weather forecasts and, more importantly, by amateurs themselves.

Water Vapor Images

Ice, water, and vapor all interact with specific bands of infrared energy; so by using suitably designed sensors in weather satellites, the satellites can detect water vapor in the atmosphere as well as in clouds. These sensors show us regions of atmospheric water vapor concentrated between altitudes of 3 and 7 km (the troposphere). Amazingly, these regions often resemble gigantic swirls and can be seen moving within the larger weather systems (see Fig. 10.5).

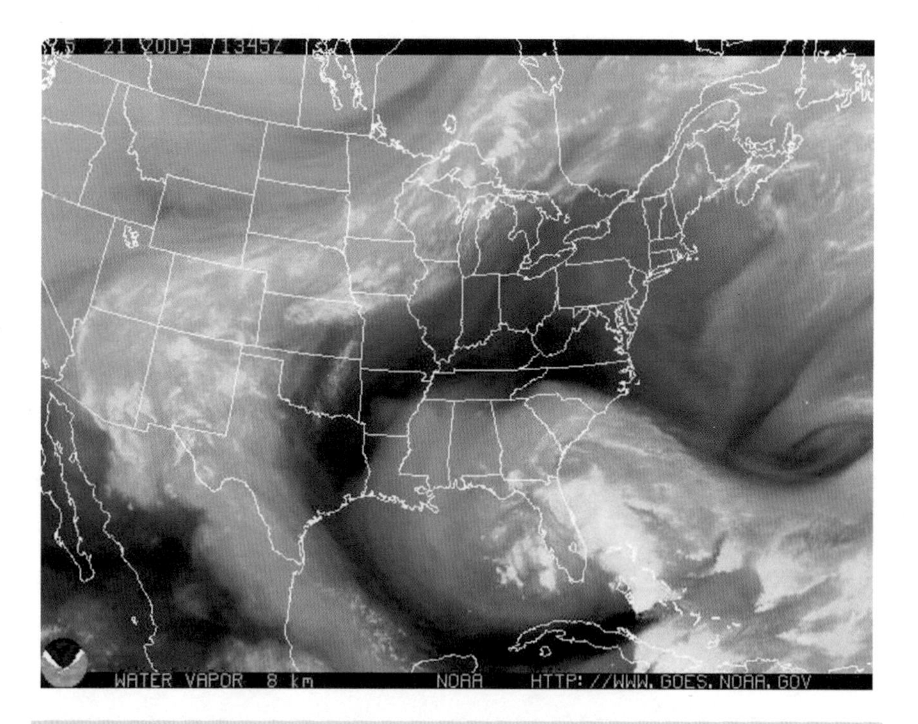

Fig. 10.5 Infrared (water vapor) LRIT image from GOES-E at 1345GMT May 21, 2009. (Image courtesy of NOAA)

Polar Orbiting Weather Satellites

The polar orbiting weather satellites are in near-circular, low Earth orbits, some 800–1,200 km in altitude. All have high orbital inclinations – the angle between their orbital planes and Earth's equator – close to 90°, so they pass near the poles on every orbit. Because Earth is rotating, they also pass over every place on Earth during each 24-h period. A study of these orbits shows that the satellite spends more time in sunshine than in darkness and travels in opposite directions at those times.

There are different types of polar orbit – notably the Sun-synchronous and the precessing types (see shortly). In addition, there are some oceanographic satellites (such as Sich and Okean) that have previously used standard a.p.t. format transmissions but have somewhat lower orbits.

The NOAA satellites are Sun-synchronous, so their orbits maintain a constant relationship with the Sun. What does this really mean? Just as geostationary satellites are at a distance where they orbit Earth once each day – and therefore keep pace with Earth's rotation – so there are lower Earth orbits that have different, but equally useful, characteristics. If a weather satellite passes over a given place at about the same time each day (for example, mid-day), transmitted images will be produced under near constant conditions of illumination. There are different operational NOAA satellites passing over an area at various times during the day; so if you use a satellite tracking program (such as an online program) to produce a list of pass times for any NOAA satellite during a period of several days, you can see that any particular satellite always passes at its maximum elevation during the same part of the day – approximately 1900GMT for NOAA-19. In this instance, the orbit's ascending node is the time at which the satellite crosses the equator traveling north and will therefore perform a south to north pass over the local horizon a few minutes later. Earlier and later passes of the same satellite will be at lower maximum elevations.

Table 10.1 shows some of the currently operating weather satellites, including the European METOP-A. The latest state of the operational a.p.t. (and other) satellites can be monitored using Yahoo's weather satellite reports forum:

http://tech.groups.yahoo.com/group/weather-satellite-reports/

From monitoring records by my own high resolution transmission (h.r.p.t. – see later) equipment, I used to report weekly on transmissions information until we

Table 10.1 Currently operational polar orbiting weather satellites

Satellite	METOP-A	NOAA-15	NOAA-16	NOAA-17	NOAA-18	NOAA-19
Ascending node	2130	1655	1712	2143	1339	1400
Status	AM primary	AM secondary	PM secondary	AM backup	PM primary	Operational verification
APT	No	Yes	No	Yes	Yes	Yes

were ready to move house, requiring the sale of the equipment. However, most polar satellites also transmit a.p.t., which can be received on low-cost equipment. In Table 10.1, AM means morning, and PM means afternoon passes. Each sequence is always followed about 12 h later by the satellite traveling in the opposite direction. After the passage of time, even Sun-synchronous satellite orbital planes drift from their original position. Due to the long-term effects of solar activity and Earth's gravitational anomalies, all satellite orbits are subjected to perturbations. Some satellites carry propulsion systems that can be used to counteract these perturbations. Others, notably the Russian meteors, are in orbits that precess – that is, their orbital planes slowly move around Earth with respect to the Sun.

NOAA Polar Orbiters

The satellites shown in Table 10.1 as transmitting a.p.t. are all NOAA satellites. They transmit continuous image data in two forms: high resolution picture telemetry (h.r.p.t.) on selected frequencies in the 1,700 MHz band, and low resolution imagery (a.p.t.) in the v.h.f. band on selected frequencies between 137 and 138 MHz. The NOAA a.p.t. line is half a second long and contains calibration components, the low resolution image from a visible-light channel, and one from the infrared channel – see Fig. 10.6. Consequently, each half-second line of NOAA data carries a large amount of useful information with two spectral image components, both derived by reducing the amount of real-time data from the corresponding high resolution image data for simultaneous transmission.

Fig. 10.6 A few minutes of a typical NOAA-14 pass showing infrared (*left side*) and visible (*right side*) sections on December 30, 1997. The upper end of the pass is entering twilight

Receiving and Decoding NOAA Weather Satellite Images

During the 1980s, the huge increase in the number of people following this hobby coincided with – and was significantly helped by – the extraordinarily rapid development of fast microprocessors and the consequent application of new software. One important development was the writing of software for the new, faster processors to decode the real-time signals received from a.p.t. satellites, a process previously done by expensive hardware. A few companies that were already designing receivers, and with staff who were also familiar with the increasing interest in weather satellite monitoring, were able to produce the hardware necessary to decode the signals. Firms such as Timestep Weather Systems and Dartcom were originally retailing products to professionals and amateurs who wanted to buy "off-the-shelf" equipment. There was also a keen interest by many amateurs in building their own decoding equipment.

For the satellite enthusiast, the weather satellite scene of the second half of the twentieth century had perhaps greater interest in the form of different satellites than is the situation today. You could guarantee raised eyebrows if you told your neighbors that you were receiving signals from Russian satellites! The reality was that – as explained in the earlier paragraph – the agreement to use a standard transmission format (a.p.t.) for polar orbiting weather satellites meant that equipment able

Fig. 10.7 Crossed dipole antenna for weather satellite reception. This is conveniently lightweight and easy to fit to a mast

to receive signals from the American satellites could also receive signals from comparable Russian ones!

A typical system designed to receive and decode a.p.t. images includes a suitable antenna, feed, receiver, and decoder. One of the most common types of antenna is the crossed dipole, phased for the right-circular polarization signals transmitted by NOAA satellites (see Fig. 10.7). Another popular type is the quadrifilar helix antenna, which can sometimes provide better quality reception than the crossed dipole. The signal is fed to a specially designed weather satellite receiver using low-loss, balanced cable; without these, possible interference problems affecting the 137 MHz band are likely to be experienced. Conventional scanners with their wide bandwidth are not normally able to produce a signal of sufficient quality to ensure good reception for decoding. The most common method of decoding these a.p.t. weather satellite signals now involves using a computer running one of the various specialist programs designed to convert the audio signals into a fully processed image in real-time (see Fig. 10.8). The good news is that the software is mostly free.

Fig. 10.8 A dedicated laptop receiving a.p.t. from a nearby external antenna. The laptop does not require a high specification

Hardware and Software

http://www.geo-web.org.uk/shop.html
The Group for Earth Observation's website – see below – is one site offering the hardware for receiving satellite data:

http://www.hffax.de/html/hauptteil_wxsat.htm
Wxsat by Christian Bock is a freeware software sound-card program that is comprehensive:

http://www.wxtoimg.com/
WXtoImg by Craig Anderson, the basic software program, is freeware and more than adequate for sound-card decoding:

http://www.poes-weather.com/
APTdecoder by Patrik Tast is a freeware program with many features:

Other useful software such as satellite-tracking programs is also available as freeware, with a few websites also offering local predictions. If you leave a dedicated computer receiving and decoding imagery, then such software may not be considered essential. The final result will be a sequence of images received every few hours from one of the NOAA satellites. During the late afternoon and early evening, the infrared image will show the latest situation exactly as the weather forecasters see it – except that you will receive it possibly a little earlier than they do! In most circumstances, it will probably give strong indications of the weather – particularly the cloud – situation for the next few hours. Overnight, the visible light channel of NOAA satellites normally switches over to an infrared channel based on the water vapor part of the spectrum – comparable with that seen in Fig. 10.5 from the GOES satellite.

Settled conditions are particularly easy to interpret. Rapidly moving weather fronts can be seen approaching or leaving, and their associated cloud systems can be correlated with the clouds that you see above. Consequently, even with nothing more than a sequence of polar orbiter images, you can judge likely observing prospects for the next few hours.

Websites with Current a.p.t. Imagery

A number of weather satellite enthusiasts have automated their image reception for Internet display, enabling their latest results to be seen at any time. Sometimes, such sites can disappear overnight if circumstances change.

http://www.wxsat.org/wxtoimg.htm
Rob Denton is an Englishman now living in Bulgaria, and he has an automated reception and publishing system that shows the latest a.p.t. from all currently operational satellites, as seen from his location in eastern Europe.

http://wx.prettygoodprojects.com/apt/daily/
Peter Goodhall provides a comprehensive display, including associated imagery with alternative processing techniques, as received in Oxford, UK.

http://www.heymoe1.com/satellite.html
Mike Optie of Elgin, Illinois, provides another comprehensive a.p.t. satellite display. His site includes several of the associated processing alternatives such as precipitation estimates. Additionally, Mike offers a table of forthcoming passes.

http://www.nigelheasman.com/wxtoimg/wxtoimg.htm
Nigel Heasman receives a.p.t. at Kayalar in northern Cyprus and also posts both imagery and forthcoming passes as recorded there.

http://www.gyweather.com/wxsatpic.php
Tony Le Page from St Peter Port, Guernsey, in the Channel Islands includes the latest composite images together with routinely posted satellite a.p.t. As with most monitors, images are provided as clickable thumbnails in order to provide a large number of satellite images for easy viewing.

http://www.weather.daleh.id.au/
Dale Hardy operates his receiving equipment at Marks Point NSW in Australia and posts a.p.t. images throughout the day.

http://www.g4iat.co.uk/wx/noaa/
Finally, Barry Smith of Darwen, UK, is one of a small number of a.p.t. monitors covering Britain continuously.

For a list of other a.p.t. image sites as well as the latest images as received by Patrik Tast in Finland, visit his site (see address in previous software section) and select "weblink," followed by "Daily APT Log" (in that order).

There are also numerous Russian early experimental and more recent oceanographic weather satellites, but discussion of them would stray too far from our main subject of helping the astronomer, so now we progress to the geostationary satellites that can help us so much.

Geostationary Weather Satellites: Image Formats

As discussed earlier, there is a constellation of geostationary weather satellites positioned along the belt at various longitudes, of which the United States has two satellites – GOES-East and GOES-West – to cover its extended longitude span. Figure 10.3 illustrates the basic constellation comprising the two GOES satellites that are accompanied in orbit by a Japanese satellite, a Russian one, and a European satellite.

During recent decades, new satellites have enabled older ones to be reassigned as backups, and – in the case of some of the European Meteosat satellites – used for entirely new projects. The main transmission formats used for weather satellite imagery are HRIT and LRIT, although other formats (such as GVAR) are also used by specific agencies, and in some cases, the main formats are not transmitted.

HRIT (high resolution) images require the largest of dishes for direct reception, and the most expensive equipment, far beyond the average amateur's funding capabilities.

The Low Rate Information Transmission (LRIT) service is the new digital data broadcast standard being implemented on recent geostationary meteorological satellites for transmission to the next generation of relatively low-cost user stations. It has replaced the earlier, very long-lived analog Wefax standard for image data as well, now providing some other geostationary meteorological satellite transmissions. The new standard has been agreed to by the Coordination Group for Meteorological Satellites (CGMS) for implementation worldwide by its members as they update their current systems.

In the following listing – see List 10.1 – HRIT refers to high resolution data for which a very large dish is required (and is therefore not normally received directly

List 10.1. Location of operational geostationary weather satellites. (Note that Europe is making use of at least two operational satellites, Meteosat-8 and 9.)

Meteosat-8 (9.5°East)
Data: HRIT (digital) and LRIT (digital)
Operational: Yes, RSS and Standby for Meteosat-9

Meteosat-9 (0°)
Data: HRIT (digital) and LRIT (digital)
Operational: Yes

GOES-12 (75.0° West)
Data: GVAR (digital) – 1,685.7 MHz
LRIT (digital): 1,691 MHz
Operational: Yes

GOES-11 (135° West)
Data: GVAR (digital) – 1,685.7 MHz
LRIT (digital): 1,691 MHz
Operational: Yes

GOES-10 (60°West)
GVAR (digital): 1,685.7 MHz
LRIT: No service

Meteosat-7 (57.5°East)
HRI (digital): 1,691 MHz
Wefax (analog) occasionally for ranging purposes only
Operational: Yes

MTSAT-1R (140° East)
Data: HRIT (digital) and LRIT(digital) on 1,691 MHz
Operational: Yes

Feng Yun-2C (105° East)
Data: SVISSR (digital) and LRIT (digital)
Operational: Yes

Feng Yun-2D (86.5° East).
Data: SVISSR (digital) and LRIT (digital)
Operational: Yes. Parallel operation with 2C.

Table 10.2 GOES spacecraft orbital locations

Satellite	GOES-8	GOES-9	GOES-10	GOES-11	GOES-12	GOES-13
Location	195° W	200° W	60° W	135° W	75° W	105° W
Status	Deactivated	Deactivated	South America Support	West Primary	East Primary	Backup

by amateurs); LRIT refers to the low resolution data that can be received by professionals or well-equipped amateurs. RSS refers to the Rapid Scanning Service provided by Eumetsat's EumetCast (see shortly). GVAR is GOES Variable Format data. See Figs. 10.2 and 10.4 for typical LRIT images.

List 10.1 and Table 10.2 show the orbital slots allocated to current and recent GOES weather satellites. The NOAA Geostationary Operational Environmental Satellite (GOES) system is for short-range warning and "now-casting," with the Polar-Orbiting Environmental Satellites (POES) for longer-term forecasting. Both kinds of satellites are necessary for providing a complete global weather monitoring system. The satellites carry search and rescue instruments and have helped save the lives of about 10,000 people to date. The satellites are also used to support aviation safety (volcanic ash detection), and maritime/shipping safety (ice monitoring and prediction).

Figure 10.2 (see the beginning of this chapter) shows an LRIT visible light image received from GOES-12, the satellite positioned over the east coast of the United States. This is just a portion of the whole disc image that is available.

GEONETCast: A Worldwide System

Figure 10.3 and Table 10.2 illustrate the spread of geostationary weather satellites, although amateurs cannot necessarily pick up low cost transmissions from them all. GEONETCast is a relatively new system, led by three regional infrastructure providers: EUMETSAT in Europe provides the EUMETCast service, the Chinese Meteorological Administration (CMA) in the Asia-Pacific region provides (or will provide) FengYunCast, and NOAA in the western hemisphere provides GEONETCast-Americas. (Feng Yun translates as "Wind and Clouds.") GEONETCast satellite coverage is expanding across the world, but the individual services are not necessarily identical!

GEONETCast receiving technology is based on the idea of using widespread, off-the-shelf components, thereby allowing for the widespread adoption of the service at low cost. The official claim is that an entire receiving station can be purchased and installed for $2,000–$3,000 (see website below), though in Britain, the cost can be considerably less. GEONETCast-Americas serves North, Central, and South America, begun in early 2008 using inexpensive satellite receiver stations based on digital video broadcast standards.

How does GEONETCast work? Each of the providers contributes its own data, whether image or pure data streams, on a scheduled basis. EUMETCast – see next section – is an excellent example of this. Eumetsat – the European organization that plans, launches, and operates the European Meteosat Second Generation (MSG) weather satellites – receives the direct transmissions of HRIT (high resolution images) and LRIT (low resolution images) from Meteosat-9 (its formal name is Meteosat Second Generation-2) and processes this data in house (as NOAA does with GOES data). Eumetsat then retransmits this processed data in real time to another satellite (HotBird), which then retransmits it on a television band frequency (approximately 12 GHz) as EumetCast data. Registered users can receive this transmission on low-cost reception equipment. This immediately makes the highest quality imagery (HRIT) easily available to European amateurs, including suitably equipped amateur astronomers!

The GEONETCast website is http://www.geonetcastamericas.noaa.gov/.

A huge variety of image and other data is transmitted from EumetCast at about 12 GHz (the same band used by terrestrial television services) using small dishes between 60 and 100 cm (depending on your location within the transmission footprint of HotBird), together with low-cost receivers. There are a few additional requirements, such as a registered dongle (a small device that slots into your computer to permit decryption) and bespoke software, but none of the associated items for EumetCast reception is pricey, and that includes the addition of animating software such as that published and updated frequently by David Taylor (see website below).

Figure 10.9 shows a typical EumetCast system required for receiving the digital video broadcasts. A standard television satellite receiving dish and suitable low

Fig. 10.9 EumetCast DVB reception equipment for Europe's MSG-2 imagery broadcast from the HotBird satellite

noise block (LNB) operating in the 12 GHz band of frequencies is pointed at HotBird-6 for European data. Correct identification of the satellite transponder is performed using a prepared instruction set made available by the data provider, with further operator advice provided by experienced amateurs. European users are provided with a dongle to permit decryption when they register to use the data. Once the signal from HotBird has been located, the sequence of software setting up and decoding can be performed. The whole process is likely to take about half a day (maybe more, but probably less), and the end result is spectacular HRIT imagery, among other weather-related imagery that may also be of interest. For the astronomer, the HRIT images are in a class of their own, coming every 15 min and calling out to be animated!

Essential software is available at nominal cost, and many users also use David Taylor's award-winning *MSG Data Manager* and *MSG Animator* programs (see website for links). Europe has been providing this low-cost entry to some of the highest quality continuous imagery for some years.

David Taylor's website is http://www.satsignal.net/. On this site, select "Using EumetCast", from which a complete description can be read about equipment and software requirements.

Animating Geostationary Weather Satellite Images

However you obtain them, whether by direct reception of the satellite data, via EumetCast within Europe, or from a source on the web, by their very nature the images from GOES and Meteosat can be animated. Each successive image shows exactly the same geographical coverage as the previous image, so software can run a sequence of these images.

For the best results, up-to-date imagery is virtually essential. Successive infrared images of the same region show the gradual movement of cloud systems regardless of the time of day or night. It means that if you are under cloud, not only can its full extent be seen but an estimate of its likely duration can also be made. You can soon judge whether it is worth waiting up for another hour or so to await the clearance. Similarly, if you are imaging under clear skies, it becomes easy to see whether a transient cloud is the start of something more extensive or is virtually on its own. There are many occasions when you might be able to use the satellite imagery to judge whether you should wait for a cloud bank to pass or finish when the cloud front arrives in an hour or so.

Some Features Seen in Satellite Imagery

The large-scale views provided by geostationary weather satellites ensure that the earliest signs of hurricane formation are usually detected on satellite images. They mostly occur over the large expanses of oceans. Low-pressure systems have a characteristic cloud structure and can be seen at mid-latitudes. Shower clouds have an easily recognizable structure, appearing as individual small blobs, often moving fairly quickly. With experience, many weather patterns can be recognized and a basic forecasting skill acquired. This is how the astronomer can quickly learn to spot near-term weather.

Although as astronomers we do not need to delve deeper into the tremendous amount of data that is available to professional meteorologists, by using these images you can understand how the professional can use satellite imagery to identify cloud shapes, heights, and types. By applying the fastest computers yet developed, weather forecasters can identify what is happening now and what is most probably going to happen to the weather in the near term and not too far future. All three types of image – visible light, infrared, and water vapor – complement each another. Weather features are often clearly seen in the individual bands that are difficult to see in the others.

Here now are the websites we referred to earlier:

http://www.dartcom.co.uk/home/index.php
http://www.eumetsat.int/Home/index.htm

Although offering an extremely valuable resource to the avid astronomer, potentially saving many wasted hours of waiting for clouds to clear that never do or missing out on clear skies that we had not expected, such satellite systems are not for everyone. If and when GEOnetCast comes your way, you are at least assured of a low-cost alternative.

For the final chapter, we will look at some of the many projects undertaken by amateur astronomers all over the world. These demonstrate the main reason that so many people take up this hobby.

Chapter 11

Some LX200 and LX400 Projects

Having completed the procedures described in this book, your telescope should now be optimized and functioning extremely well. With a CCD camera fitted to it you can expect to be able to obtain good-quality images even without autoguiding, at least for moderately short exposures – perhaps up to 2 min, though 90 s may be more normal for some telescopes. The universe should now be open to your scope.

Your original decision to buy one of the Meade LX200GPS or LX400 series of telescopes was likely based on a wish to involve yourself in serious astronomy, or perhaps simply raise your chances of taking some really impressive deep space images. This chapter provides some insight into the projects being carried out by amateur and professional astronomers using these telescopes.

Lunar Impact Studies

During past decades, many amateur astronomers studying the Moon using larger telescopes – perhaps above 8 in. (20 cm) in diameter – have occasionally reported seeing flashes of light on the dark areas of the Moon. Enough reports existed for the idea to be taken seriously, although no professional study was made until fairly recently. Since 2005, NASA astronomers have been observing the Moon with a purpose. According to team leader Rob Suggs of the Marshall Space Flight Center the monitoring program was started in 2006 after NASA announced plans to return astronauts to the Moon. If astronauts were going back, it was important to know whether there was any real danger from impacts.

In May 2006, the first instrument used was a 10 in. (25 cm) Newtonian reflecting telescope having fast optics (f/4.7), enabling the NASA astronomers to cover about one-sixth of the visible Moon (or one-third of the dark side). In June 2006, a second instrument, a 14 in. (35 cm) LX400 telescope with a focal reducer – see Fig. 11.1 – was

L. Harris, *So You Want a Meade LX Telescope!*, Patrick Moore's Practical Astronomy Series, DOI 10.1007/978-1-4419-1775-1_11, © Springer Science+Business Media, LLC 2010

Fig. 11.1 A 14 in. (35 cm) LX400 telescope with a focal reducer and astronomical video cameras at the Automated Lunar and Meteor Observatory (ALaMO). This telescope is used to observe the Moon for lunar impact flashes (NASA) (Images credit Science@NASA)

located in one of the observatory domes. The 25 cm telescope was replaced by another 35 cm telescope in early September 2006. Further facilities were built to enhance the program, allowing simultaneous observations of the Moon with three identical 35 cm telescopes. This helped to eliminate cosmic ray camera hits on one observing system being misidentified as lunar meteoroid flashes.

A flash was detected on November 7, 2005, when a piece of Comet Encke hit Mare Imbrium. The resulting explosion produced a magnitude 7 flash – an easy target for the team's 25 cm telescope. According to Bill Cooke, head of NASA's Meteoroid Environment Office at the Marshall Space Flight Center, when meteoroids hit the Moon, a typical blast is about as powerful as a few hundred pounds of TNT and can be photographed easily using a backyard telescope. Figure 11.2 shows individual video frames of the May 2, 2006, lunar impact flash as observed from ALaMO. In January 2008, one of the 35 cm telescopes was replaced with a 50 cm f/8.1 telescope from Ritchey–Chrétien Optical Systems.

Every year in early January, the Earth–Moon system passes through a stream of debris from Comet Quadrantid. Although on Earth, Quadrantids disintegrate as

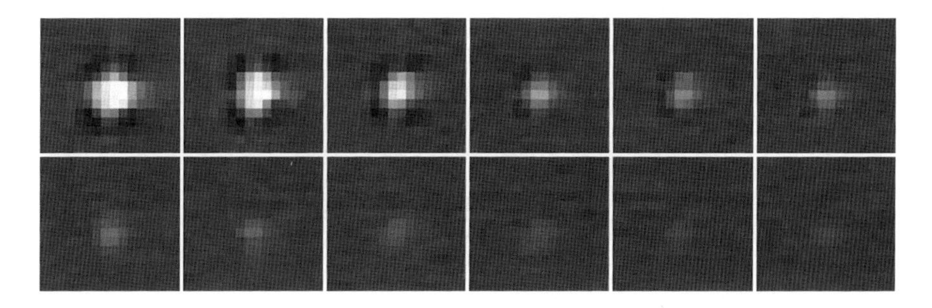

Fig. 11.2 Example flash image sequence. (Credit NASA Images Marshall Space Flight Center)

flashes of light in the atmosphere, on the airless Moon they hit the ground and explode. During meteor showers, the Moon passes through comet debris, causing the rate of lunar flashes to increase as high as one per hour. Interestingly, the rate never goes as low as zero even during nonmeteor shower periods. See the website for details showing where lunar impacts have been recorded.

(http://www/nasa.gov/centers/marshall/news/lunar/photos.html
http://www.nasa.gov/centers/marshall/news/lunar/)

With NASA using LX400 and comparable telescopes for studying the incidence of lunar impacts, the amateur astronomer can utilize a 25 or 30 cm telescope for use on those occasions when the Moon's phase provides scope (excuse the pun) for such monitoring. Note that this is usually performed using a web camera to record the monitoring results, rather than just observing the Moon visually all the time. The LX200GPS and LX400 telescopes offer a lunar rate drive for convenience when looking at the Moon.

Searching for Extrasolar Planets

Another highly advanced project for the LX200GPS/LX400 telescope owner is the searching and monitoring of exoplanets, or extrasolar planets, as they are known. Circa 2009 there are more than 300 such planets known to orbit distant stars. Of these, some 43 have been observed by amateurs. Nicoladj Haarup of Silkeborg, Denmark, works with his 12 in. (30 cm) LX400, and on February 17, 2008, Nicoladj observed XO-2b (see Fig. 11.3), making it the first exoplanet observation from Danish ground. His inspiration came partly from the book *Exoplanet Observing for Amateurs* written by Bruce L. Gary. The principle of observing exoplanets involves making careful measurements of the target star magnitude before, during, and after a transit. An amateur can contribute real science with LX200GPS or LX400 telescopes.

For further information and resources for exoplanet monitoring, visit the following URL: http://brucegary.net/AXA/x.htm

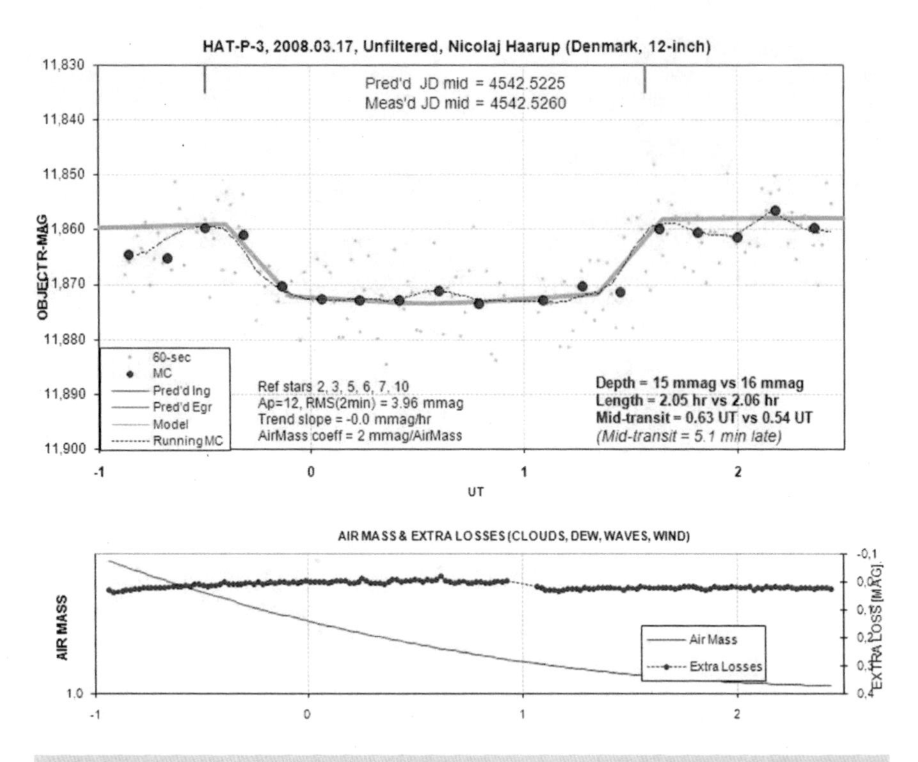

Fig. 11.3 Exoplanet observation project showing XO-2b (Image courtesy Nicoladj Haarup; http://www.starworks.dk/index.php)

Comet Imaging

The author's standard exposure time has been 10 min, but you can take much shorter exposures of comets with no guiding. This is because comets are, of course, moving relative to the stars, so unless the comet is bright enough to directly guide on, you are building in a smearing of the nucleus. With some interlaced cameras (those with pixel readouts along alternate rows) it is possible to actually guide on a comet using the imaging camera itself, although you lose 50% of the data. You might prefer to take shorter multiple exposures, knowing that LX200GPS and LX400 telescopes can cope with unguided runs of up to about 120 s.

You might enjoy imaging an occasional comet, at least the ones that escape the notice of fellow astronomers! Programs such as *Guide*-8 can provide a list – see Table 11.1 – of currently visible comets, appropriate to your observation point. Note that Table 11.1 is a much reduced version of the actual results. A comparable list of asteroids can also be produced. The list includes estimated magnitudes and visibility (whether above or below the horizon at the selected time, as well as whether evening or morning comets). From this, you can list those comets – often more than one – that you will try to image that evening. This goes into your "jobs

Table 11.1 Edited Guide-8 listing showing currently visible comets down to magnitude 15.0

Name (designation)	Mag	RA	Dec	Constellation	Elongation
P/Boethin (85P)	10.4	03h28m19.23s	+23 38′ 58.5″	Ari	82.5 Ev
Broughton (C/2006 OF2)	12.2	06h04m15.12s	+43 31′ 43.0″	Aur	115.9 Ev
Cardinal (C/2008 T2)	10.4	03h14m32.80s	+61 56′ 05.1″	Cas	91.6 Ev
P/Christensen (210P)	14.0	15h56m13.72s	–02 01′ 12.5″	Ser	95.5 Mo
Christensen (C/2006 W3)	13.4	22h26m36.84s	+35 38′ 58.9″	Lac	46.0 Ev
P/Christensen (P/2008 X4)	14.0	15h56m13.68s	–02 01′ 10.2″	Ser	95.5 Mo
P/Churyumov-Gerasimenko (67P)	13.1	01h16m25.19s	+08 44′ 08.5″	Psc	47.8 Ev
P/Gunn (65P)	14.5	12h37m52.20s	+10 50′ 10.9″	Vir	146.2 Mo
P/Kearns-Kwee (59P)	14.5	03h08m25.71s	+25 48′ 48.9″	Ari	78.8 Ev

for that session" file (a list of all targets) and includes the optimum time for imaging. On one particular night in February 2009, Comet Kushida was well placed. Had the sky remained clear (it was due to cloud over within about 3 h), then Comet Lulin would have been the priority.

The process was as follows. The comet was identified as being suitably bright and in a favorable position using *Guide-8* updated with recent comet elements from the Minor Planet Center. The telescope had already been synchronized with the sky after taking a 10-s exposure and using *PinPoint LE* to solve the image. By entering the coordinates of the comet into *MaxIm's* coordinate entry location option and selecting *goto*, the scope drove to the comet. A sample image showed the comet to be a little off center, but because the image scale was known to the software (following an earlier image scale calibration), the *point the telescope here* option was used – clicking the mouse at the center of the comet. This time, the short exposure image check showed the comet at center. The exposure sequence was set to take multiple 100s exposures with the one-shot color camera and the observatory did its own thing.

Monitoring of the observatory computer operations is easily carried out using the remote desktop connection option in *Windows XP Pro*, so one possible procedure is to have a second computer ready to connect from indoors. This facility allows one to monitor and control telescope operations. Your system may be limited if you cannot control rotation of the dome and therefore must frequently pop outside to check and rotate the dome and shutter whenever necessary. For this particular sequence of exposures no interaction was required. In the above case the planned sequence of eight 100s exposures of the comet was obtained without problems.

If the camera has very low dark current it need not be normal practice to take darks although 10-min darks every few months are advisable for long term monitoring of dark noise. If you do not normally do dark frame subtraction, hot pixels can be a problem. All hot pixels remain in the same position on each image, so you can load a sample image and create a new hot pixel map in *MaxIm DL* using the *Process, remove bad pixels* option.

A new map can be created as follows: change the display by right clicking on the image and selecting *low*. This improves the contrast, raising the relative brightness of the hot pixels. By clicking carefully on each hot pixel, its position is recorded. A map of pixel positions is built up and then *saved*. Reselect *remove bad pixels* and *process*. Any remaining hot pixels that were inadvertently wrongly selected can be added to the map and the new map resaved. Use *process* again, and if necessary repeat until there are no more noticeable hot pixels. You can then run a *batch process* to remove hot pixels from each individual raw image, and then convert each image to color. This whole sequence takes minutes to complete, and the map should remain valid for subsequent raw images. Of course, cosmic ray hits will not be removed; this is done in a separate process explained in image processing books.

Finally, use the *combine files* command to combine the collection of converted color images. You can sum all the comet images using the one-star click synchronization method in which you click the mouse on the center of the comet in each successive image. The resulting image shows the comet itself against a background of trailed stars – see Fig. 11.4. You can select star synchronization and produce an image showing the slow movement of the comet's nucleus across the starry background; it is all a matter of taste.

At the time of imaging, Comet Kushida was 0.84 AU (Astronomical Units – 1 AU being the average Earth-to-Sun distance). Another successful short comet imaging sequence was that of c2006 OF2 (Broughton), though only four images were obtained due to the onset of high mist or cloud terminating the evening's short session.

Fig. 11.4 Final 13-min exposure of Comet Kushida using fixed nucleus position

Astrometry

A very large number of LX200GPS and LX400 scopes are all but dedicated to astrometry. As discussed briefly in Chap. 8 (Advanced Software), this is the measuring of the precise positions of heavenly bodies – often asteroids and comets. *Astrometrica* and *Visual PinPoint* are two of the various programs able to perform this task. They, and all comparable programs, require a database of stars down to very low magnitudes. You can use the *USNO A2.0* catalog, though more recent catalogs are becoming available, often having stars with improved positional accuracy. If you have never tried astrometry then we can recommend that you at least have a try at measuring the positions of some of the brighter asteroids and comets. The basic method has been briefly discussed in Chap. 8, but it is now appropriate to include some more details.

Although it is possible to use planetarium software such as *Guide-8* on its own, with the asteroids and comet files fully updated, *The Sky* and *Starry Night Pro Plus* in conjunction with *ACP Planner* can provide an efficient means of doing this. One method involves measuring only a dozen or so asteroids, using *Guide-8* after updating the asteroid database. Using the tables option *Guide* can produce a list of asteroids down to any selected limiting magnitude. This list can be sorted under any criterion – such as via magnitude. For your first measurements, you should opt for a selection of the brighter asteroids – perhaps magnitudes 14–16. From the analysis of previous short exposure (say 60 s) images, you should be familiar with your limiting magnitude for various sky conditions. You can hope to see down to magnitude 17 or beyond in a clear sky in suburbia assuming there is no mist at higher levels. Consequently, measuring this limiting magnitude should be the *first* task of any evening. If the limiting magnitude stalls at about 16 (or even less), then you know that at least for that time, astrometry could be severely limited. In Chap. 10, we gave details about how you can keep virtually completely up to date with the real weather situation at your telescope.

Meanwhile, back to the asteroid list. After having established which Right Ascension bands are easily visible (taking due note of associated declinations as well), you can quickly identify a selection of suitable asteroids for astrometry during that time and include a comet list as well – these being potentially more exciting due to the possibility of spotting an unexpected brightening. You might wish to produce comparable lists for later hour bands during the night. Clearly this process would be far more efficiently performed using planning software, but you might not wish to dedicate a whole session to astrometry.

With the prepared asteroid/comet lists, you can obtain the orbital elements from the Minor Planet Center as explained in Chap. 8. Perhaps use *ACP* to run the plan and produce the first set of measurements. After an elapse of at least 30 min for each asteroid, rerun the plan a second time, and later a third time to obtain the three images required for each asteroid position measurement. Figure 8.2 in Chap. 8 shows the result of a typical detection run.

Deep Sky Pictures

In earlier chapters, we discussed the methods of improving the tracking of your LX200GPS or LX400 telescope, and one measure of your success will be the duration of the longest exposure that you can make that will still produce circular stars in most of a number of consecutive images. With a fully optimized telescope, you are able to quickly assess its capabilities for deep sky imaging. Using a suitable analytical tool, such as *Astrometrica*, you can analyze occasional images to check on the faintest detectable star and therefore plan an observation session accordingly.

Some of the author's own examples, but it must be noted that the deep sky images credited to him have not undergone any significant image processing. The aim is to show what can be expected as a *minimum*. By using a selection of the numerous image processing techniques available in advanced software packages, one can bring out the best from one's own images. Think of these images as a starting point!

Monitoring the LX200 groups found on Yahoo and elsewhere reveals members' archived and other images of both Solar System and deep space images. Such images are often the result of multiple long exposures and extended image processing. Following my request on the LX200GPS and RXC400 forums, a number of astronomers kindly provided me with access to their images, some of which are reproduced here with appropriate credits. Many of the images have been processed further by the contributors to achieve the best result. A variety of images has been selected (a) some images that have involved no guiding at all, (b) some with normal autoguiding, and (c) some using narrowband filters. Contributors are listed in alphabetical order of surname. Please note that space limitations prevented me including more.

Andrey Batchvarov

Using an 8 in. (20 cm) LX200 operating at f6.3 Andrey Batchvarov has taken a large number of deep sky images from which I am reproducing M8, a well-known Messier object. In both cases, Andrey has processed multiple 1-min unguided images to produce Fig. 11.5 taken with a Canon 300D camera. The image illustrates how well deep sky pictures can be produced by adding multiple exposures.

George Hall

Using his 12 in. (30 cm) LX200GPS at f10, an ST-8XME guide camera with AO-8 adaptive optics unit guiding at 10 Hz for luminance and from 1 to 4 Hz for the red, green, and blue filters, George Hall imaged M1 – see Fig. 11.6 – from his driveway in Dallas, Texas. Exposures included 24 luminance 5-min subframes, and six 5-min subframes using RGB filters. Images were subsequently processed using darks and flats in *CCDStack* and *Photoshop.*

Fig. 11.5 M8 8 in. (20 cm) LX200 SCT operating at f6.3, Canon 300D camera, image processing *FizFix*, 21×1min unguided exposures (Image credit Andrey Batchvarov; http://aida.astroinfo.org/index.php?cat=10008)

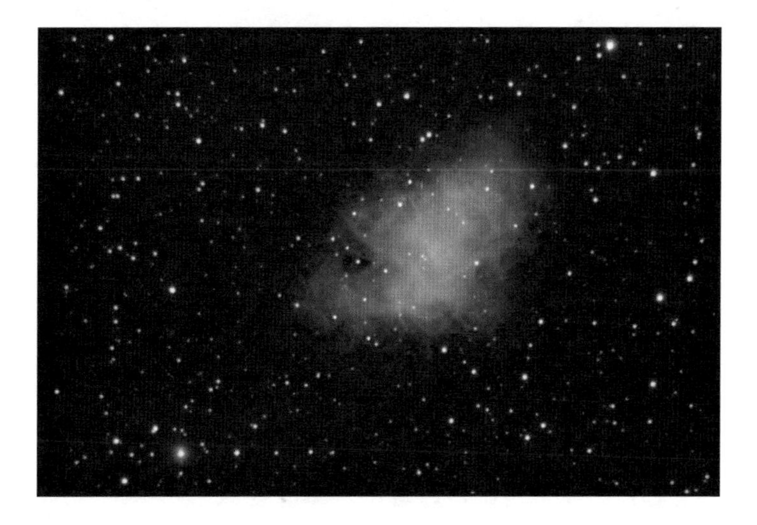

Fig. 11.6 M1 taken with 12 in. (30 cm) LX200GPS using ST-8XME guide camera and SBIG AO-8 adaptive optics unit (Image credit George Hall; http://homepage.mac.com/joanlvh/Astronomy/Equipment.html)

Hilary Jones

Using his Meade 8-in. (20 cm) LX200GPS scope and DSI-C camera, Hilary has taken many deep space images of which Fig. 11.7 is an example. The exposure totals 205 min for M51 and comprises numerous separate frames, some using filters.

Fig. 11.7 M51 (Courtesy Hilary Jones; http://www.muffycat.org/astronomy/)

Fig. 11.8 M16 taken June 2, 2006 (Courtesy Bill Norby)

Bill Norby

Using his 10-in. (25 cm) Meade LX200 Classic, an ST-2000XCM camera with AO-7 adaptive optics unit for autoguiding, Bill Norby of St. Peters, Missouri, took 79 two-minute exposures (total 2 h 38 min) of M16 (the Eagle Nebula) – see Fig. 11.8. After applying dark frame and flat field corrections, he processed and produced his final image using *CCDSoft v5*.

Richard Robinson

Figure 11.9, which shows NGC7293 (the Helix Nebula), was taken by Richard Robinson using a 8 in. (20 cm) f10 LX200 fitted with the Meade f3.3 focal reducer. The 80 min total exposure with a Starlight Xpress MX-5C and STAR2000 tracking resulted in four 20-min exposures, which were combined in *AstroArt*.

Stuart Thompson

This image of the Eta Carina Nebula (Fig. 11.10) comes from Stuart Thompson. Using a Meade 10-in. (25 cm) LX400 telescope operating at f/6, with an Astro-Physics 0.75 focal reducer. Stuart used a set of narrowband filters (Astronomik Ha, SII, and OIII filters) for luminance, blue, green, and red components, each exposed for 30 min in 5-min subframes, and with an SBIG ST – 10XME camera with an SBIG AO-8 adaptive optics unit.

Fig. 11.9 NGC7293 total exposure 80 min using 8-in. (20 cm) LX200 (Image credit Richard Robinson; http://wordepot.com/RAO/index.html)

Fig. 11.10 A 10-in. (25 cm) LX400 telescope operating at f6 of Eta Carina Nebula (Image credit Stuart Thompson; http://sites.google.com/site/antipodeanastrophotography/)

Fig. 11.11 NGC7635 (Bubble Nebula) (Image credit Charles Trump; http://www.ct-computerservice.com/Astro.htm)

Charles Trump

This 2-h combined color filtered image of NGC7635 – see Fig. 11.11 – was taken by Charles Trump using a 12-in. (30 cm) LX400 telescope. *Maxlm DL* was used for image capture and subsequent processing.

Merope and Its Nebulosity

With so many superb deep space images taken professionally – such as by the Hubble Space Telescope – and by amateurs prepared to spend hours on one object, we are almost overwhelmed with choices for targets. After seeing an amateur's impressive image of a region of the Pleiades cluster, the author decided to try to capture the nebulosity surrounding Merope. With an adaptive optics unit fitted to the scope behind the f6.3 converter, Merope was added to a list of long exposure targets planetarium program *Guide-8* showed the time of transit, and therefore the optimum time for the middle of the exposure sequence. To obtain the best possible result, a set of twilight flats was taken. As described in a previous chapter, these are essentially designed to remove the effects of any vignetting and dust motes. Using the f6.3 reducer does not produce significant vignetting in my scope, but for long exposures, dust doughnuts can spoil the images of any bright object, especially where there is background nebulosity.

Twilight Flats

Without a light box (a box that fits over the end of the scope and is designed to provide a uniformly illuminated surface from which a set of flat images can be produced at any time), twilight flats must be made. This is done by taking very short exposure daylight images of the sky high and well away from the sunset region, as the Sun sets below the horizon. Very short exposures taken late in the day can usually produce an image that has a uniform background and does not saturate. By scaling the monitor screen and displaying pixel information for the whole frame, a gradual reduction in the average pixel level will be seen as twilight approaches.

A good rule of thumb based on the recommendation of several image processing experts is to wait until the average level is somewhat below half maximum; one school of thought suggests about 30%. With a maximum pixel level of 65,535, aim to obtain flats with the average level at around 25–30,000. This region is well within the linear response of the CCD chip and is therefore likely to produce good data. Always produce at least ten flats and use John Winfield's *Maxlm* plug-in called *SkyFlats* (see URL below). This superb piece of free software is a small add-in program that *Maxlm DL* can invoke for this purpose. When the sky is approaching the right brightness level (use the *information* window to monitor it continuously), launch the plug-in with the latest sample flat open. You should enter the initial exposure time based on your test settings, and the target ADU count. Other settings should be appropriate for your own computer.

John Winfield's URL: http://winfij.homeip.net/maximdl/index.html

After running the flat collecting session (which takes about 1 min following the *settings* checking), run a de-Bayer process on each flat if you are using a one-shot-color camera. (These OSC cameras have a Bayer matrix of color filters over each pixel; the de-Bayer filter effectively removes this for our purposes.) The results are then combined to produce a master flat. You can use the *MaxIm DL stack* command and the *sigma clip* option. This works well with the ten flats, extracting the most representative pixel from each flat to produce the master flat. Interestingly, if you leave the scope not tracking, the sigma-clip process also eliminates any odd stars inadvertently captured from the frames. It should be pointed out here that a rigorous processing sequence could also include flat and dark adjustment and bias subtraction, but a discussion of this is getting beyond the scope of this book. Read a more detailed description of other processing options in the book *Creating and Enhancing Digital Astro Images* by Grant Privett (Springer).

Processing

If you are using a mono camera, color conversion can obviously be omitted. The individual raw 10-min images should undergo hot pixel removal as a first stage. This was discussed previously and can be completed as a batch process. You should then convert the raw images into floating point format using *Batch save and convert*. This ensures that the color conversion (or subsequent process) – which is the next stage – can be completed without any inadvertent loss of data when the maximum ADU (camera sensor level count) exceeds 65,535.

To remove any scope artifacts use the *pixel math* option to divide the converted image by the master flat. This process can also be done using the batch process window because it is performed on each image. To avoid confusion, set the new images to be saved in a new folder, or alternatively be saved with a new name. The result can be a set of color images all flat converted. The final stage involves using the stack command to combine all the color images using *align*, *auto star matching*, and *sigma clip*. This latter option is only valid if you used a dither to obtain your multiple images. The *dither* command ensures that each image is slightly offset from the other images; the combination strongly rejects hot pixels. It is still preferable to remove hot pixels from the original raw images before finally combining them, but the process works well whichever way you choose. Figure 11.12 shows the final image of Merope in M45 including the surrounding nebulosity. This is an 80-min total exposure comprising eight 10-min subframes.

Two other images – Figs. 11.13 and 11.14 – were taken using my usual telescope and camera combination, the 12-in. (30 cm) LX400 and SXV-H9C one-shot color camera.

Fig. 11.12 Merope in M45 showing surrounding nebulosity. It is an 80-min exposure comprising eight 10-min subframes (Image by the author)

Fig. 11.13 NGC772, 70 min total exposure (Image by the author)

Fig. 11.14 IC5146, 120 min total exposure using a 12-in. LX400 scope and SXV-H9C camera (Image by the author)

Each image comprises 10-min subframes at full resolution, using the Starlight Xpress adaptive optics unit and an average guide exposure of between 0.1 and 0.5 s. As indicated earlier, *no* processing other than basic hot pixel removal and flat correction has been done.

Author's Note

I am extremely grateful to the chapter reviewers who freely loaned me their well-known expertise to help ensure that silly oversights were trapped before submission. I had wanted to include pictures from Meade's factory showing the production of these telescopes, but Meade declined to provide any.

The completion of this book took place between house moves. We moved from Southampton, UK, to Stowmarket in Suffolk, with a delayed final move pending this completion. Our new property has a good-sized garden for my observatory to be rehoused. I shall therefore once more be going through every stage described in this book to again optimize the telescope's performance. Hopefully I might eventually achieve the one ambition that I have not fulfilled – as mentioned in an earlier chapter. Did you guess what it is? My first discovery! Perhaps you will beat me to it. Or perhaps you already have!

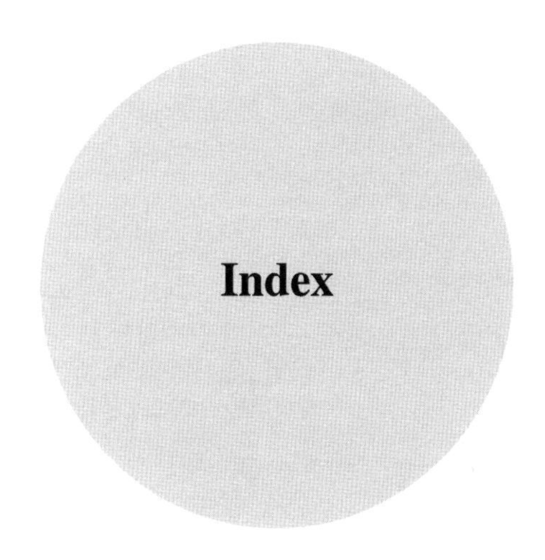

Index